Encyclopaedia

of

OPTICAL SCIENCE, TECHNOLOGY AND ENGINEERING

Encyclopaedia of Optical Science, Technology and Engineering

Prof. G.K. Bose

ANMOL PUBLICATIONS PVT. LTD.
NEW DELHI-110 002 (INDIA)

ANMOL PUBLICATIONS PVT. LTD.

Regd. Office: 4360/4, Ansari Road, Daryaganj,
New Delhi-110 002 (India)
Ph.: 23278000, 23261597

Branch Office: No. 1015, Ist Main Road, BSK IIIrd Stage
IIIrd Phase, IIIrd Block,
Bangalore-560 085 (India)
Tel.: 080-41723429
Visit us at: www.anmolpublications.com

Encyclopaedia of Optical Science, Technology and Engineering

First Edition, 2009

ISBN 978-81-261-4130-2

PRINTED IN INDIA

Printed at Balaji Offset, Delhi.

Contents

Preface

Optics is the science that describes the behavior and properties of light and the interaction of light with matter. Optics explains optical phenomena. The word *optics* means *appearance* or *look.* Optics usually describes the behavior of visible, infrared, and ultraviolet light; however because light is an electromagnetic wave, similar phenomena occur in X-rays, microwaves, radio waves, and other forms of electromagnetic radiation and analogous phenomena occur with charged particle beams. Optics can largely be regarded as a sub-field of electromagnetism. Some optical phenomena depend on the quantum nature of light relating some areas of optics to quantum mechanics. In practice, the vast majority of optical phenomena can be accounted for using the electromagnetic description of light, as described by Maxwell's equations.

The field of optics has its own identity, societies, and conferences. The pure science aspects of the field are often called optical science or optical physics. Applied optical sciences are often called optical engineering. Applications of optical engineering related specifically to illumination systems are called illumination engineering. Each of these disciplines tends to be quite different in its applications, technical skills, focus, and professional affiliations. More recent innovations in optical engineering are often categorized as photonics or optoelectronics. The boundaries between these fields and "optics" are often unclear, and the terms are used differently in different parts of the world and in different areas of industry.

Author

Chapter 1

Nature of Light

WAVE NATURE

An age-old debate which has persisted among scientists is related to the question, "Is light a wave or a stream of particles?" Very noteworthy and distinguished physicists have taken up each side of the argument, providing a wealth of evidence for each side. The fact is that light exhibits behaviors which are characteristic of both waves and particles.

Light exhibits certain behaviors which are characteristic of any wave and would be difficult to explain with a purely particle-view.

Light reflects in the same manner that any wave would reflect. Light refracts in the same manner that any wave would refract. Light diffracts in the same manner that any wave would diffract. Light undergoes interference in the same manner that any wave would interfere.

And light exhibits the Doppler effect just as any wave would exhibit the Doppler effect. Light behaves in a way that is consistent with our conceptual and mathematical understanding of waves. Since light behaves like a wave, one would have good reason to believe that it might be a wave. We will investigate the variety of behaviors, properties and characteristics of light which seem to support the wave model of light. Here we will focus on three specific behaviors - reflection, refraction and diffraction.

A wave doesn't just *stop* when it reaches the end of the medium. Rather, a wave will undergo certain behaviors when it encounters the end of the medium. Specifically, there will be some reflection off the boundary and some transmission into the new medium. The transmitted wave undergoes refraction (or bending) if it approaches the boundary at an angle.

If the boundary is merely an obstacle implanted within the medium, and if the dimensions of the obstacle are smaller than the wavelength of the wave, then there will be very noticeable diffraction of the wave around the object. Each one of these behaviors - reflection, refraction and diffraction - is characterized by specific conceptual principles and mathematical equations. Now we will see how light waves demonstrate their wave nature by reflection, refraction and diffraction.

REFLECTION OF LIGHT WAVES

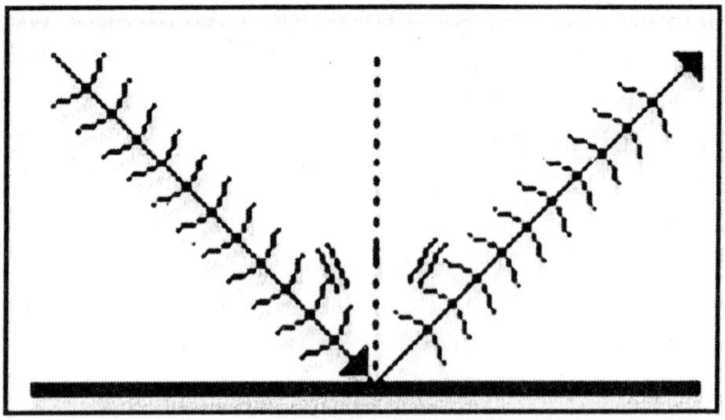

Fig. Light Waves Follow the "Law of Wave Reflection."

All waves are known to undergo reflection or the bouncing off of an obstacle. Most people are very accustomed to the fact that light waves also undergo reflection. The reflection of light waves off of a mirrored surface results in the formation of an image.

One characteristic of wave reflection is that the angle at which the wave approaches a flat reflecting surface is equal to the angle at which the wave leaves the surface. This characteristic is observed for water waves and sound waves. It is also observed for light waves. Light, like any wave, follows the law of reflection when bouncing off surfaces.

REFRACTION OF LIGHT WAVES

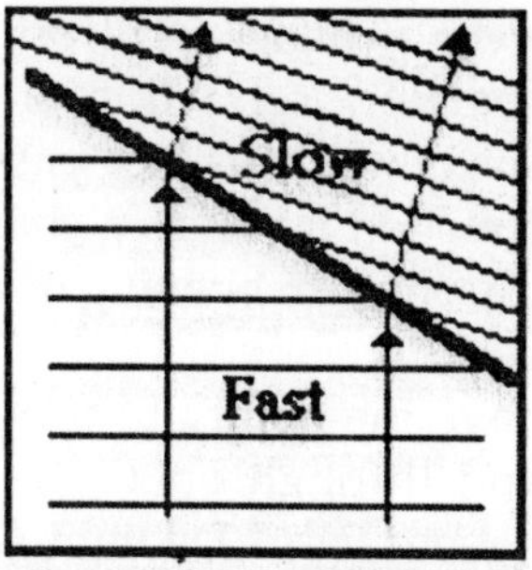

Fig. The Refraction Behaviour of Light
Reveals that light is very Wave - like

All waves are known to undergo refraction when they pass from one medium to another medium. That is, when a wavefront crosses the boundary between two media, the direction that the wavefront is moving undergoes a sudden change; the path is "bent."

This behaviour of wave refraction can be described by both conceptual and mathematical principles. First, the direction of "bending" is dependent upon the relative speed of the two media. A wave will bend one way when it passes from a medium in which it travels slow into a medium in which it travels fast; and if moving from a *fast medium* to a *slow medium*, the wavefront will bend in the opposite direction.

Second, the amount of bending is dependent upon the actual speeds of the two media on each side of the boundary. The amount of bending is a measurable behaviour which follows distinct mathematical equations. These equations are based upon the speeds of the wave in the two media and the angles at which the wave approaches and departs from the boundary.

Light, like any wave, is known to refract as it passes from one medium into another medium. In fact, a study of the refraction of light reveals that its refractive behaviour follows the same conceptual and mathematical rules which govern the refractive behaviour of other waves such as water waves and sound waves.

DIFFRACTION OF LIGHT WAVES

Reflection involves a change in direction of waves when they bounce off a barrier. Refraction of waves involves a change in the direction of waves as they pass from one medium to another. And diffraction involves a change in direction of waves as they pass through an opening or around an obstacle in their path.

Water waves have the ability to travel around corners, around obstacles and through openings. Sound waves do the same. But what about light? Do light waves bend around obstacles and through openings? If they do, then it would provide still more evidence to support the belief that light behaves as a wave.

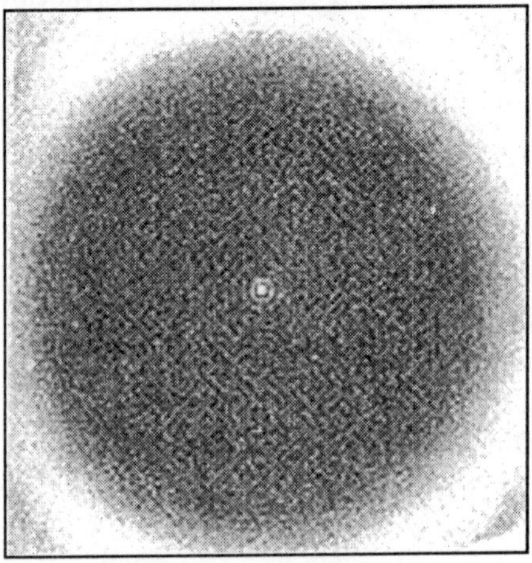

Fig. Light Diffracts around a penny-observe the light spot in the centre of pattern

When light encounters an obstacle in its path, the obstacle blocks the light and tends to cause the formation of a shadow in the region behind the obstacle. Light does not exhibit a very noticeable ability to bend around the obstacle and fill in the region behind it with light. Nonetheless, light does diffract around obstacles.

In fact, if we observe a shadow carefully, we will notice that its edges are extremely fuzzy. Interference effects occur due to the diffraction of light around different sides of the object, causing the shadow of the object to be fuzzy. This is often demonstrated in a Physics classroom with a laser light and penny demonstration. Light diffracting around the right edge of a penny can constructively and destructively interfere with light diffracting around the left edge of the penny.

The result is that an interference pattern is created; the pattern consists of alternating rings of light and darkness. Such a pattern is only noticeable if a narrow beam of monochromatic light (i.e., single wavelength light) is passed directed at the penny. The photograph at the right shows an interference pattern created in this manner.

Since, light waves are diffracting around the edges of the penny, the waves are broken up into different wavefronts which converge at a point on a screen to produce the interference pattern shown in the photograph. This amazing penny diffraction demonstration provides another reason why believing that light has a wavelike nature makes cents.

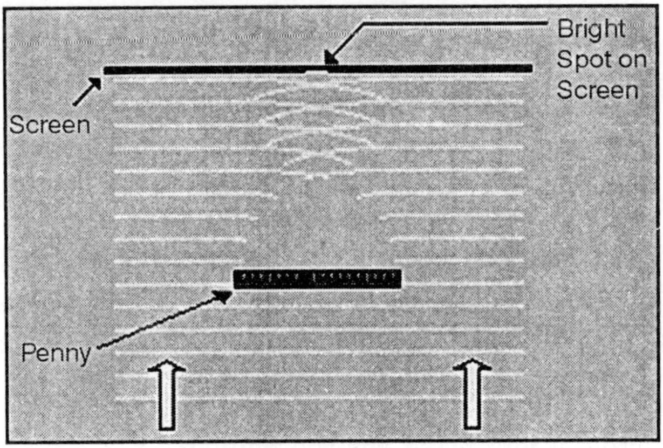

Fig. The Diffraction of light around a penny and its ultimate interference upon a screen produces a pattern which could only be observed if light is a wave.

Light behaves as a wave - it undergoes reflection, refraction, and diffraction just like any wave would.

TWO POINT SOURCE INTERFERENCE

Wave interference is a phenomenon which occurs when

two waves meet while traveling along the same medium. The interference of waves causes the medium to take on a shape which results from the net effect of the two individual waves upon the particles of the medium. Wave interference can be constructive or destructive in nature.

Constructive interference occurs at any location along the medium where the two interfering waves have a displacement in the same direction. For example, if at a given instant in time and location along the medium, the crest of one wave meets the crest of a second wave, they will interfere in such a manner as to produce a "super-crest." Similarly, the interference of a trough and a trough interfere constructively to produce a "super-trough."

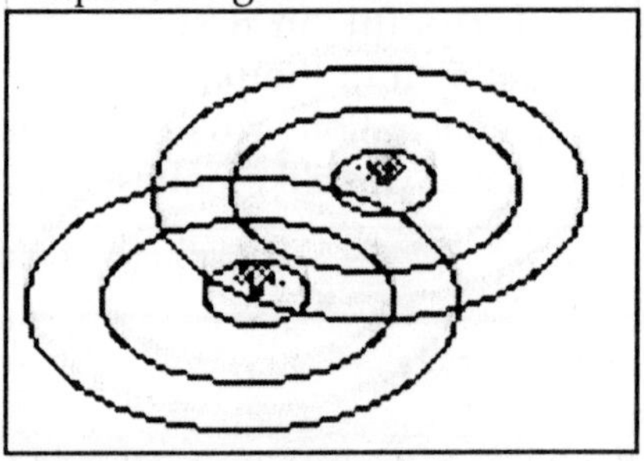

Fig. Two Simultaneous Disturbances in a Pond will Create Circular Waves which Interfere as they Spread over the Surface of the Pond

Destructive interference occurs at any location along the medium where the two interfering waves have a displacement in the opposite direction. For example, the interference of a crest with a trough is an example of destructive interference. Destructive interference has the tendency to decrease the resulting amount of displacement of the medium.

A defining moment in the history of the debate concerning the nature of light occurred in the early years of the nineteenth century.

Thomas Young showed that an interference pattern results when light from two sources meets up while traveling through the same medium. To understand Young's experiment, it is important to back up a few steps and discuss the interference of water waves which originate from two points.

The value of a ripple tank in the study of water wave behaviour was introduced and discussed. If an object bobs up and down in the water, a series water waves in the shape of concentric circles will be produced within the water. If two objects bob up and down with the same frequency at two different points, then two sets of concentric circular waves will be produced on the surface of the water.

These concentric waves will interfere with each other as they travel across the surface of the water. If we have ever simultaneously tossed two pebbles into a lake (or somehow simultaneously disturbed the lake in two locations), we undoubtedly noticed the interference of these waves.

The crest of one wave will interfere constructively with the crest of the second wave to produce a large upward displacement. And the trough of one wave will interfere constructively with the trough of the second wave to produce a large downward displacement.

And finally the crest of one wave will interfere destructively with the trough of the second wave to produce no displacement. In a ripple tank, this constructive and destructive interference can be easily controlled and observed. It represents a basic wave behaviour which can be expected of any type of wave.

Two-Point Source Interference Patterns

The interference of two sets of periodic and concentric waves with the same frequency produces an interesting pattern in a ripple tank.

The diagram at the right depicts an interference pattern produced by two periodic disturbances. The crests are denoted by the thick lines and the troughs are denoted by the thin lines. Thus, constructive interference occurs wherever a thick line meets a thick line or a thin line meets a thin line; this type of interference results in the formation of an antinode.

The antinodes are denoted by a red dot. Destructive interference occurs wherever a thick line meets a thin line;

this type of interference results in the formation of a node. The nodes are denoted by a blue dot.

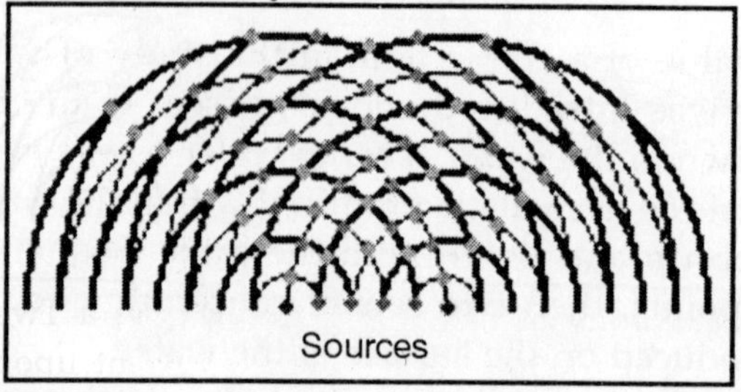

Fig. Two-Point Source Interference Pattern

The pattern is a standing wave pattern, characterized by the presence of nodes and antinodes which are "standing still" - i.e., always located at the same position on the medium. The antinodes (points where the waves always interfere constructively) seem to be located along lines - creatively called antinodal lines.

The nodes also fall along lines - called nodal lines. The two-point source interference pattern is characterized by a pattern of alternating nodal and antinodal lines. There is a central line in the pattern - the line which bisects the line segment which is drawn between the two sources is an antinodal line.

This *central antinodal line* is a line of points where the waves from each source always reinforce each other by means of constructive interference. The nodal and antinodal lines are included on the diagram below.

A two-point source interference pattern always has an alternating pattern of nodal and antinodal lines. There are however some features of the pattern which can be modified. First, a change in wavelength (or frequency) of the source will alter the number of lines in the pattern and alter the proximity or closeness of the lines.

An increase in frequency will result in more lines per centimeter and a smaller distance between each consecutive line. And a decrease in frequency will result in less lines per centimeter and a greater distance between each consecutive line.

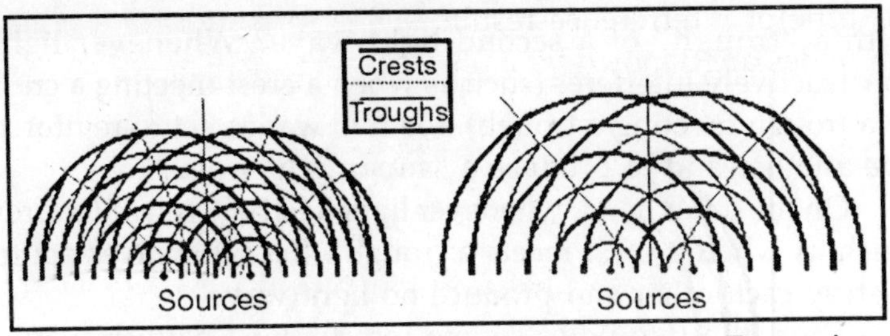

Fig. The Proximity of the Anti-nodal Lines in a Two-point Source Interference Pattern is Dependent upon the Wavelenght of the Waves.

Second, a change in the distance between the two sources will also alter the number of lines and the proximity or closeness of the lines. When the sources are moved further apart, there are more lines produced per centimeter and the lines move closer together. These two general cause-effect relationships apply to any two-point source interference pattern, whether it be due to water waves, sound waves, or any other type of wave.

Two-Point Source Light Interference Patterns

Any type of wave, whether it be a water wave or a sound wave should produce a two-point source interference pattern if the two sources periodically disturb the medium at the same frequency. Such a pattern is always characterized by a pattern of alternating nodal and antinodal lines.

Of course, the question should arise and indeed did arise in the early nineteenth century: Can light produce a two-point source interference pattern? If light is found to produce such a pattern, then it will provide more evidence in support of the wavelike nature of light.

Before we investigate the evidence in detail, let's discuss what one might observe if light were to undergo two-point source interference. What would happen if a "crest" of one light wave interfered with a "crest" of a second light wave? And what would happen if a "trough" of one light wave interfered with a "trough" of a second light wave? And finally, what would happen if a "crest" of one light wave interfered with a "trough" of a second light wave? Whenever light constructively interferes (such as when a crest meeting a crest or a trough meeting a trough), the two waves act to reinforce one another and to produce a "super light wave."

On the other hand, whenever light destructively interferes (such as when a crest meets a trough), the two waves act to destroy each other and produce no light wave.

Thus, the two-point source interference pattern would still consist of an alternating pattern of antinodal lines and nodal lines. However for light waves, the antinodal lines are equivalent to bright lines and the nodal lines are equivalent to dark lines.

If such an interference pattern could be created by two light sources and projected onto a screen, then there ought to be an alternating pattern of dark and bright bands on the screen. And since the central line in such a pattern is an antinodal line, the central band on the screen ought to be a bright band.

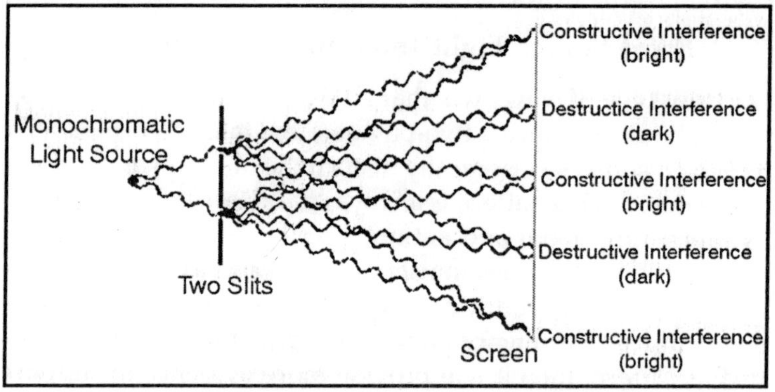

Fig. A two-point Source Interference Pattern Creates an Alternating Pattern of Bright and Dark Lines when it is Projected onto a Screen.

In 1801, Thomas Young successfully showed that light does produce a two-point source interference pattern. In order to produce such a pattern, monochromatic light must be used. Monochromatic light is light of a single color; by use of such light, the two sources will vibrate with the same frequency. It is also important that the two light waves be vibrating in phase with each other; that is, the crest of one wave must be produced at the same precise time as the crest of the second wave. (This is often referred to as coherent light.)

To accomplish this, Thomas Young used a single light source and projected the light onto two pinholes. The light from the source will then diffract through the pinholes and the pattern can be projected onto a screen. Since there is only one source of light, the set of two waves which emanate from the pinholes will be in phase with each other.

As expected, the use of a monochromatic light source and pinholes to generate in-phase light waves resulted in a pattern of alternating bright and dark bands on the screen. A typical appearance of the pattern is shown below.

Fig. The Projection of an Interference Pattern onto a Screen Results in an alternating Pattern of Bright (anti-nodes) and Dark (nodes) Bands.

Young's two-point source interference experiment is often performed in a Physics course with laser light. It is found that the same principles which apply to water waves in a ripple tank also apply to light waves in the experiment.

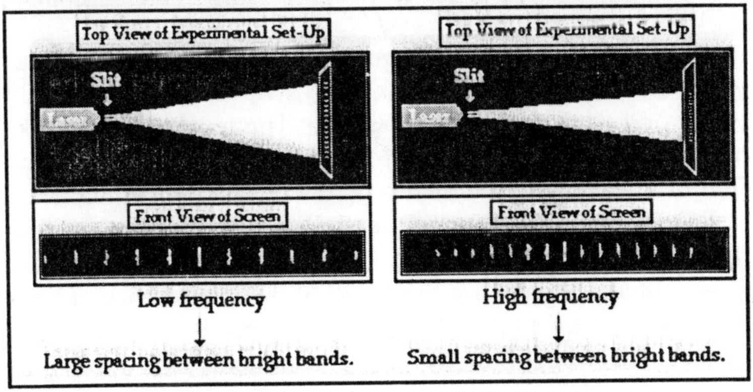

Fig. Results of Young's Experiment

For instance, a higher frequency light source should produce an interference pattern with more lines per centimeter in the pattern and a smaller spacing between lines. Indeed this is observed to be the case.

Furthermore, a greater distance between slits should produce an interference pattern with more lines per centimeter in the pattern and a smaller spacing between lines. Again, this is observed to be the case.

THIN FILM INTERFERENCE

The emphasis of this unit is to present some evidence which has historically supported the view that light behaves as a wave. The reflection, refraction and diffraction of light waves is one strand of evidence. The interference of light waves is a second strand of evidence. In the early nineteenth century, Thomas Young showed that the interference of light passing through two slits produces an interference pattern when projected on a screen.

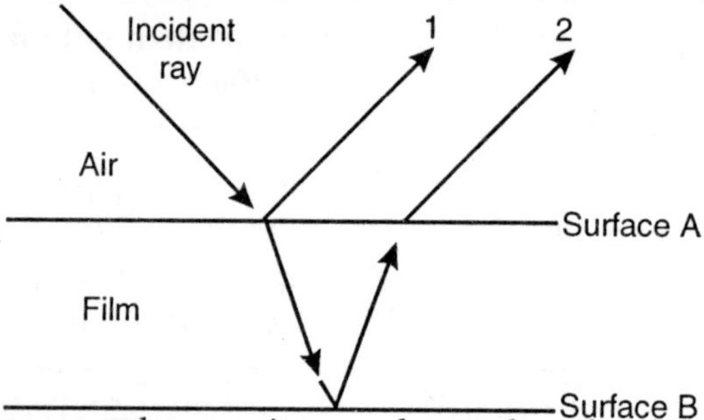

Perhaps we have witnessed streaks of color on a car windshield shortly after it has been swiped by a windshield wiper or a squeegee at a gas station. The momentary streaks of color are the result of interference of light by the very thin film of water or soap which remains on the windshield.

Or perhaps we have witnessed streaks of color in a thin film of oil resting upon a water puddle or concrete driveway. These streaks of color are the result of the interference of light by the very thin film of oil which is spread over the water surface. This form of interference is commonly called thin film interference and provides another line of evidence for the wave behaviour of light. Light wave interference results when two waves are traveling through a medium and meet up at the same location.

So what exactly is causing this thin film interference? What is the source of the two waves? When a wave (light waves included) reaches the boundary between two medium, a portion of the wave reflects off the boundary and a portion is transmitted across the boundary. The reflected portion of the wave remains in the original medium.

The transmitted portion of the wave enters the new medium and continues traveling through it until it reaches a subsequent boundary. If the new medium is a thin film, then the transmitted wave does not travel far before it reaches a new boundary and undergoes the usual reflection and transmission behaviour. Thus, there are two waves which emerge from the film - one wave which is reflected off the top of the film (wave 1 in the diagram) and the other wave which reflects off the bottom of the film (wave 2 in the diagram).

These two waves could interfere constructively if they meet two conditions. One condition is that the two waves must be relatively close together such that their crests and troughs can meet up with each other and cause the interference. To meet this condition, the light must be incident at angles close to zero with respect to the normal.

A second condition which must be met is that the wave which travels through the film and back into the original medium must have traveled just the right distance such that it is *in phase* with the other reflected wave. Two waves which are *in phase* are waves which are always at the same point on their wave cycle. That is, the two waves must be forming crests at the same location and at the same moment in time and forming troughs at the same location and at the same moment in time. In order for the second condition to occur, the thickness of the film must be *just perfect*.

If wave 1 and wave 2 meet these two conditions as they reflect and exit the film, then they will constructively interfere. Light which is visible to our eyes consists of a collection of light waves of varying wavelength. Each wavelength is characterized by its own color. So a red light wave has a different wavelength than a orange light wave which has a different wavelength than a yellow light wave. While the thickness of a film at a given location may not allow a red and an orange light wave to emerge from the film *in phase,* it may be *just perfect* to allow a yellow light wave to emerge *in phase.*

So at a given location on the film, the yellow light wave undergoes constructive interference and becomes brighter than the other colors within the incident light.

As such, the film appears yellow when viewed by incident sunlight. Other locations of the film may be just perfect to constructively reinforce red light. And still others area of the film may be of perfect thickness for the constructive reinforcement of green light. Because different locations of the film may be of appropriate thickness to reinforce different colors of light, the thin film will show streaks of color when viewed from above.

While the mathematics of thin film interference can become quite complicated, it is clear from this discussion that thin film interference is another phenomenon that can only be explained using a wave model of light.

POLARIZATION

A light wave is an electromagnetic wave which travels through the vacuum of outer space. Light waves are produced by vibrating electric charges. For our purposes, it is sufficient to merely say that an electromagnetic wave is a transverse wave which has both an electric and a magnetic component.

The transverse nature of an electromagnetic wave is quite different from any other type of wave which has been discussed. Let's suppose that we use the customary slinky to model the behaviour of an electromagnetic wave. As an electromagnetic wave traveled towards us, then we would observe the vibrations of the slinky occurring in more than one plane of vibration.

This is quite different than what we might notice if we were to look along a slinky and observe a slinky wave traveling towards we. Indeed, the coils of the slinky would be vibrating back and forth as the slinky approached; yet these vibrations would occur in a single plane of space.

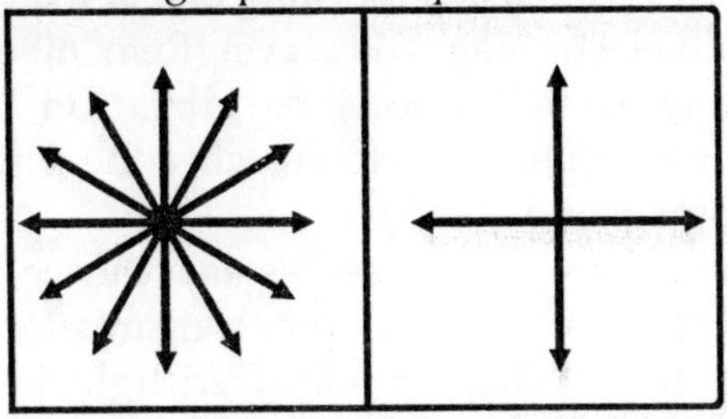

Fig. A Light Waves is known to Vibrate in a Multitude of Direction

That is, the coils of the slinky might vibrate up and down or left and right. Yet regardless of their direction of vibration, they would be moving along the same linear direction as we sighted along the slinky. If a slinky wave were an electromagnetic wave, then the vibrations of the slinky would occur in multiple planes. Unlike a usual slinky wave, the electric and magnetic vibrations of an electromagnetic wave occur in numerous planes. A light wave which is vibrating in more than one plane is referred to as unpolarized light.

Light emitted by the sun, by a lamp in the classroom, or by a candle flame is unpolarized light. Such light waves are created by electric charges which vibrate in a variety of directions, thus creating an electromagnetic wave which vibrates in a variety of directions. This concept of unpolarized light is rather difficult to visualize. In general, it is helpful to picture unpolarized light as a wave which has an average of half its vibrations in a horizontal plane and half of its vibrations in a vertical plane.

It is possible to transform unpolarized light into polarized light. Polarized light waves are light waves in which the vibrations occur in a single plane. The process of transforming unpolarized light into polarized light is known as polarization. There are a variety of methods of polarizing light.

Polarization by Use of a Polaroid Filter

The most common method of polarization involves the

use of a Polaroid filter. Polaroid filters are made of a special material which is capable of blocking one of the two planes of vibration of an electromagnetic wave. (Remember, the notion of two planes or directions of vibration is merely a simplification which helps us to visualize the wavelike nature of the electromagnetic wave.)

In this sense, a Polaroid serves as a device which filters out one-half of the vibrations upon transmission of the light through the filter. When unpolarized light is transmitted through a Polaroid filter, it emerges with one-half the intensity and with vibrations in a single plane; it emerges as polarized light.

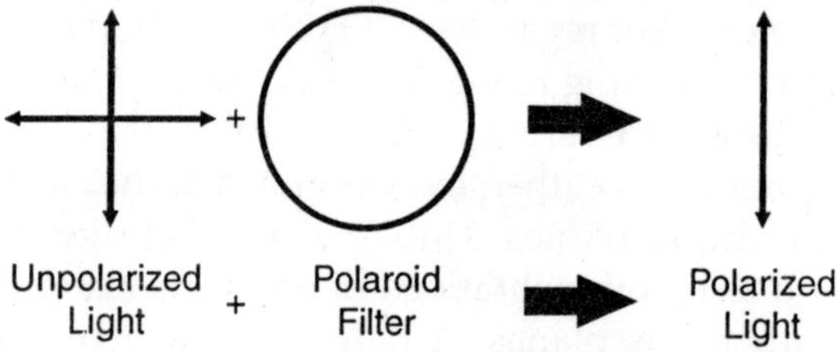

A Polaroid filter is able to polarize light because of the chemical composition of the filter material. The filter can be thought of as having long-chain molecules that are aligned within the filter in the same direction. During the fabrication of the filter, the long-chain molecules are stretched across the filter so that each molecule is (as much as possible) aligned in say the vertical direction.

As unpolarized light strikes the filter, the portion of the waves vibrating in the vertical direction are absorbed by the filter. The general rule is that the electromagnetic vibrations which are in a direction parallel to the alignment of the molecules are absorbed.

The alignment of these molecules gives the filter a polarization axis. This polarization axis extends across the length of the filter and only allows vibrations of the electromagnetic wave that are parallel to the axis to pass through. Any vibrations which are perpendicular to the polarization axis are blocked by the filter. Thus, a Polaroid filter with its long-chain molecules aligned horizontally will have a polarization axis aligned vertically. Such a filter will block all horizontal vibrations and allow the vertical vibrations to be transmitted. On the other hand, a Polaroid filter with its long-chain molecules aligned vertically will have a polarization axis aligned horizontally; this filter will block all vertical vibrations and allow the horizontal vibrations to be transmitted.

Relationship between Long-Chain Molecule Orientation and the Orientation of the Polarization Axis.

Relationship Between Long-Chain Molecular Orientation and the Orientation of the Polarization Axis

Fig. When Molecules in the Filter are Aligned Vertically, the Polarization Axis is Horizontal

Polarization of light by use of a Polaroid filter was is often demonstrated in a Physics class through a variety of demonstrations. Filters are used to look through an view objects. The filter does not distort the shape or dimensions of the object; it merely serves to produce a dimmer image of the object since one-half of the light is blocked as it passed through the filter.

A pair of filters are often placed back to back in order to view objects looking through two filters. By slowly rotating the second filter, an orientation can be found in which all the light from an object is blocked and the object can no longer be seen when viewed through two filters.

What happened? In this demonstration, the light was polarized upon passage through the first filter; perhaps only vertical vibrations were able to pass through. These vertical vibrations were then blocked by the second filter since its polarization filter is aligned in a horizontal direction. While we are unable to see the axes on the filter, we will know when the axes are aligned perpendicular to each other because with this orientation, all light is blocked.

So by use of two filters, one can completely block all of the light which is incident upon the set; this will only occur if the polarization axes are rotated such that they are perpendicular to each other.

A picket-fence analogy is often used to explain how this dual-filter demonstration works. A picket fence can act as a polarizer by transforming an unpolarized wave in a rope into a wave which vibrates in a single plane. The spaces between the pickets of the fence will allow vibrations which are parallel to the spacings to pass through while blocking any vibrations which are perpendicular to the spacings.

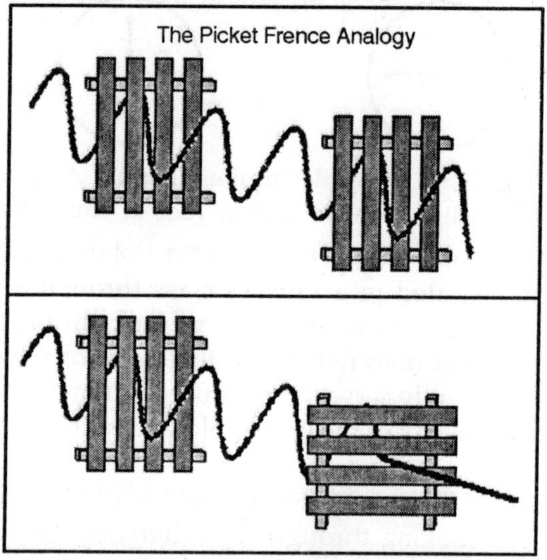

Fig. When the Pickets of both Fences are Aligned in the Vertical Direction, a Vertical Vibration can make it Through both Fences.

Obviously, a vertical vibration would not have the room to make it through a horizontal spacing. If two picket fences are oriented such that the pickets are both aligned vertically, then vertical vibrations will pass through both fences. On the other hand, if the pickets of the second fence are aligned horizontally, then the vertical vibrations which pass through the first fence will be blocked by the second fence. This is depicted in the diagram below.

In the same manner, two Polaroid filters oriented with their polarization axes perpendicular to each other will block all the light. Now that's a pretty cool observation which could never be explained by a particle view of light.

Polarization by Reflection

Unpolarized light can also undergo polarization by reflection off of nonmetallic surfaces. The extent to which polarization occurs is dependent upon the angle at which the light approaches the surface and upon the material which the surface is made of.

Metallic surfaces reflect light with a variety of vibrational directions; such reflected light is unpolarized. However, nonmetallic surfaces such as asphalt roadways, snow fields and water reflect light such that there is a large concentration of vibrations in a plane parallel to the reflecting surface. A person viewing objects by means of light reflected off of nonmetallic surfaces will often perceive a glare if the extent of polarization is large.

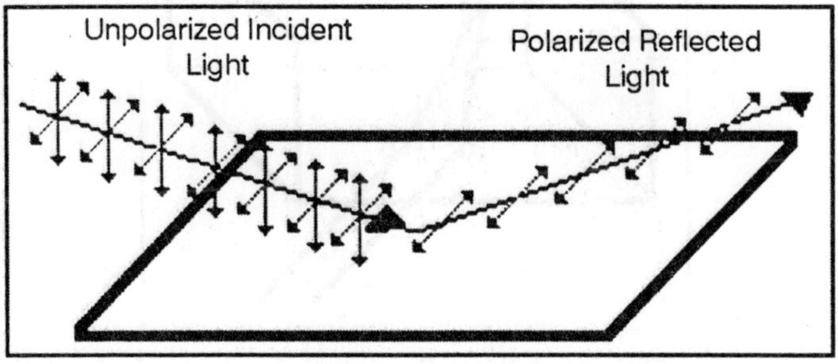

Fig. Reflection of Light off of Non-metallic Surfaces Results in some Degree of Polarization Parallel to the Surface

Fisherman are familiar with this glare since it prevents them from seeing fish which lie below the water. Light reflected off a lake is partially polarized in a direction parallel to the water's surface.

Fisherman know that the use of glare-reducing sunglasses with the proper polarization axis allows for the blocking of this partially polarized light. By blocking the plane-polarized light, the glare is reduced and the fisherman can more easily see fish located under the water.

Polarization by Refraction

Polarization can also occur by the refraction of light. Refraction occurs when a beam of light passes from one material into another material. At the surface of the two materials, the path of the beam changes its direction. The refracted beam acquires some degree of polarization. Most often, the polarization occurs in a plane perpendicular to the surface.

The polarization of refracted light is often demonstrated in a Physics class using a unique crystal which serves as a double-refracting crystal. Iceland Spar, a rather rare form of the mineral calcite, refracts incident light into two different paths. The light is *split* into two beams upon entering the crystal.

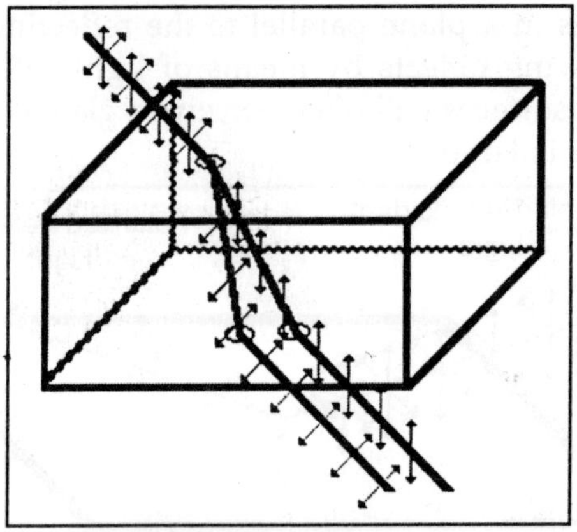

Fig. The Two Refracted Rays Passing through the Iceland Spar Crystal are Polarized with Perpendicular Orientations

Subsequently, if an object is viewed by looking through an Iceland Spar crystal, two images will be seen.

The two images are the result of the double refraction of light. Both refracted light beams are polarized - one in a direction parallel to the surface and the other in a direction perpendicular to the surface. Since these two refracted rays are polarized with a perpendicular orientation, a polarizing filter can be used to completely block one of the images.

If the polarization axis of the filter is aligned perpendicular to the plane of polarized light, the light is completely blocked by the filter; meanwhile the second image is as bright as can be. And if the filter is then turned 90-degrees in either direction, the second image reappears and the first image disappears. Now that's pretty neat observation that could never be observed if light did not exhibit any wavelike behaviour.

Polarization by Scattering

Polarization also occurs when light is scattered while traveling through a medium. When light strikes the atoms of a material, it will often set the electrons of those atoms into vibration.

The vibrating electrons then produce their own electromagnetic wave which is radiated outward in all directions. This newly generated wave strikes neighboring atoms, forcing their electrons into vibrations at the same original frequency. These vibrating electrons produce another electromagnetic wave which is once more radiated outward in all directions. This absorption and reemission of light waves causes the light to be scattered about the medium.

This scattered light is partially polarized. Polarization by scattering is observed as light passes through our atmosphere. The scattered light often produces a glare in the skies. Photographers know that this partial polarization of scattered light leads to photographs characterized by a *washed-out* sky.

The problem can easily be corrected by the use of a Polaroid filter. As the filter is rotated, the partially polarized light is blocked and the glare is reduced. The photographic secret of capturing a vivid blue sky as the backdrop of a beautiful foreground lies in the physics of polarization and Polaroid filters.

Applications of Polarization

Polarization has a wealth of other applications besides

their use in glare-reducing sunglasses. In industry, Polaroid filters are used to perform stress analysis tests on transparent plastics.

As light passes through a plastic, each color of visible light is polarized with its own orientation. If such a plastic is placed between two polarizing plates, a colorful pattern is revealed. As the top plate is turned, the color pattern changes as new colors become blocked and the formerly blocked colors are transmitted.

A common Physics demonstration involves placing a plastic protractor between two Polaroid plates and placing them on top of an overhead projector. It is known that structural stress in plastic is signified at locations where there is a large concentration of colored bands.

This location of stress is usually the location where structural failure will most likely occur. Perhaps we wish that a more careful stress analysis was performed on the plastic case of the CD which we recently purchased.

Polarization is also used in the entertainment industry to produce and show 3-D movies. Three-dimensional movies are actually two movies being shown at the same time through two projectors. The two movies are filmed from two slightly different camera locations.

Each individual movie is then projected from different sides of the audience onto a metal screen. The movies are projected through a polarizing filter. The polarizing filter used for the projector on the left may have its polarization axis aligned horizontally while the polarizing filter used for the projector on the right would have its polarization axis aligned vertically.

Consequently, there are two slightly different movies being projected onto a screen. Each movie is cast by light which is polarized with an orientation perpendicular to the other movie. The audience then wears glasses which have two Polaroid filters. Each filter has a different polarization axis - one is horizontal and the other is vertical.

The result of this arrangement of projectors and filters is that the left eye sees the movie which is projected from the right projector while the right eye sees the movie which is projected from the left projector. This gives the viewer a perception of depth.

Our model of the polarization of light provides some substantial support for the wavelike nature of light. It would be extremely difficult to explain polarization phenomenon using a particle view of light.

Polarization would only occur with a transverse wave. For this reason, polarization is one more reason why scientists believe that light exhibits wavelike behaviour.

Chapter 2

Reflection and Refraction of Light

RAYS AND WAVE FRONTS

Light is a very complex phenomenon, but in many situations its behaviour can be understood with a simple model based on rays and wave fronts. A ray is a thin beam of light that travels in a straight line.

A wave front is the line (not necessarily straight) or surface connecting all the light that left a source at the same time.

For a source like the Sun, rays radiate out in all directions; the wave fronts are spheres centered on the Sun. If the source is a long way away, the wave fronts can be treated as parallel lines.

Rays and wave fronts can generally be used to represent light when the light is interacting with objects that are much larger than the wavelength of light, which is about 500 nm. In particular, we'll use rays and wave fronts to analyse how light interacts with mirrors and lenses.

THE LAW OF REFLECTION

Objects can be seen by the light they emit, or, more often, by the light they reflect. Reflected light obeys the law of reflection, that the angle of reflection equals the angle of incidence.

For objects such as mirrors, with surfaces so smooth that any hills or valleys on the surface are smaller than the wavelength of light, the law of reflection applies on a large scale. All the light travelling in one direction and reflecting from the mirror is reflected in one direction; reflection from such objects is known as specular reflection.

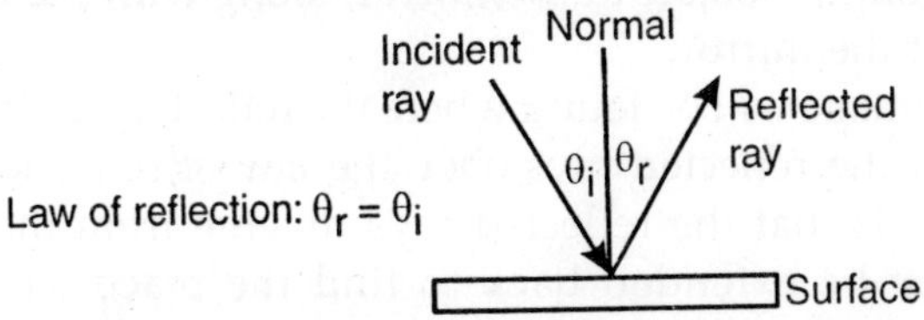

Most objects exhibit diffuse reflection, with light being reflected in all directions. All objects obey the law of reflection on a microscopic level, but if the irregularities on the surface of an object are larger than the wavelength of light, which is usually the case, the light reflects off in all directions.

PLANE MIRRORS

A plane mirror is simply a mirror with a flat surface; all of us use plane mirrors every day, so we've got plenty of experience with them. Images produced by plane mirrors have a number of properties, including:

- The image produced is upright
- The image is the same size as the object (i.e., the magnification is m = 1)
- The image is the same distance from the mirror as the object appears to be (i.e., the image distance = the object distance)
- The image is a virtual image, as opposed to a real image, because the light rays do not actually pass through the image. This also implies that an image could not be focused on a screen placed at the location where the image is.

A LITTLE GEOMETRY

Dealing with light in terms of rays is known as geometrical optics, for good reason: there is a lot of geometry involved. It's relatively straight-forward geometry, all based on similar triangles, but we should review that for a plane mirror. Consider an object placed a certain distance in front of a mirror.

To figure out where the image of this object is located, a ray diagram can be used. In a ray diagram, rays of light are drawn from the object to the mirror, along with the rays that reflect off the mirror.

The image will be found where the reflected rays intersect. Note that the reflected rays obey the law of reflection. What we notice is that the reflected rays diverge from the mirror; they must be extended back to find the place where they intersect, and that's where the image is.

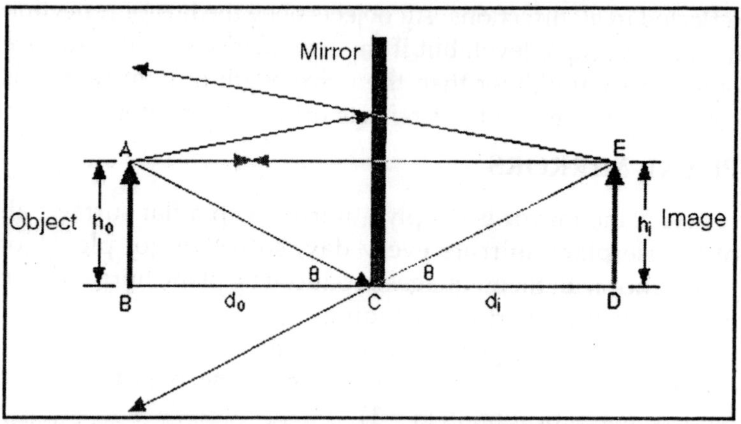

Analyzing this a little further, it's easy to see that the height of the image is the same as the height of the object. Using the similar triangles ABC and EDC, it can also be seen that the distance from the object to the mirror is the same as the distance from the image to the mirror.

SPHERICAL MIRRORS

Light reflecting off a flat mirror is one thing, but what happens when light reflects off a curved surface? We'll take a look at what happens when light reflects from a spherical mirror, because it turns out that, using reasonable approximations, this analysis is fairly straight-forward. The image we see is located either where the reflected light converges, or where the reflected light appears to diverge from.

A spherical mirror is simply a piece cut out of a reflective sphere. It has a centre of curvature, C, which corresponds to the centre of the sphere it was cut from; a radius of curvature, R, which corresponds to the radius of the sphere; and a focal point (the point where parallel light rays are focused to) which is located half the distance from the mirror to the centre of curvature. The focal length, f, is therefore:

focal length of a spherical mirror : $f = R / 2$

This is actually an approximation. Parabolic mirrors are really the only mirrors that focus parallel rays to a single focal point, but as long as the rays don't get too far from the principal axis then the equation applies for spherical mirrors. The diagram shows the principal axis, focal point (F), and centre of curvature for both a concave and convex spherical mirror.

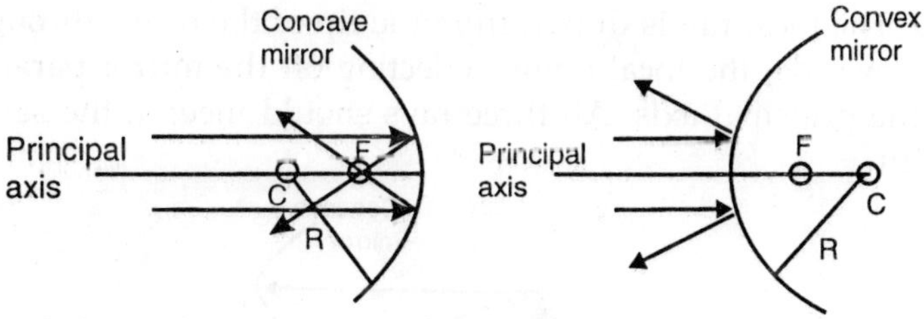

Spherical mirrors are either concave (converging) mirrors or convex (diverging) mirrors, depending on which side of the spherical surface is reflective. If the inside surface is reflective, the mirror is concave; if the outside is reflective, it's a convex mirror. Concave mirrors can form either real or virtual images, depending on where the object is relative to the focal point.

A convex mirror can only form virtual images. A real image is an image that the light rays from the object actually pass through; a virtual image is formed because the light rays can be extended back to meet at the image position, but they don't actually go through the image position.

RAY DIAGRAMS

To determine where the image is, it is very helpful to draw a ray diagram. The image will be located at the place where the rays intersect. We could just draw random rays from the object to the mirror and follow the reflected rays, but there are three rays in particular that are very easy to draw.

Only two rays are necessary to locate the image on a ray diagram, but it's useful to add the third as a check. The first is the parallel ray; it is drawn from the tip of the object parallel to the principal axis. It then reflects off the mirror and either passes through the focal point, or can be extended back to pass through the focal point.

The second ray is the chief ray. This is drawn from the tip of the object to the mirror through the centre of curvature. This ray will hit the mirror at a 90° angle, reflecting back the way it came. The chief and parallel rays meet at the tip of the image.

The third ray, the focal ray, is a mirror image of the parallel ray. The focal ray is drawn from the tip of the object through (or towards) the focal point, reflecting off the mirror parallel to the principal axis. All three rays should meet at the same point.

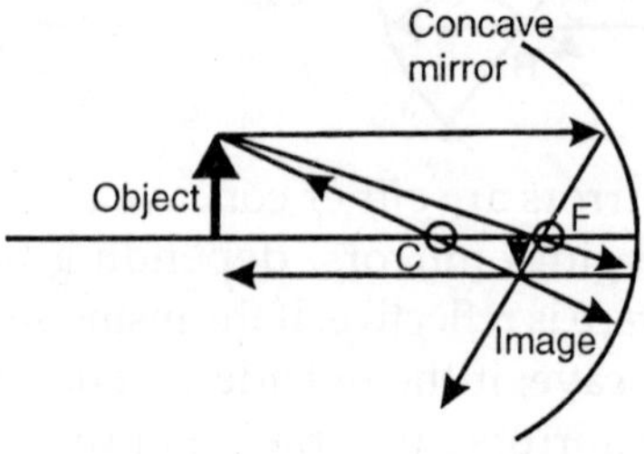

Fig. A Ray Diagram for a Concave Mirror

This shows a few different things. For this object, located beyond the centre of curvature from the mirror, the image lies between the focal point (F) and the centre of curvature. The image is inverted compared to the object, and it is also a real image, because the light rays actually pass through the point where the image is located.

With a concave mirror, any object beyond C will always have an image that is real, inverted compared to the object, and between F and C. We can always trade the object and image places (that just reverses all the arrows on the ray diagram), so any object placed between F and C will have an image that is real, inverted, and beyond C.

What happens when the object is between F and the mirror? We should draw the ray diagram to convince ourself that the image will be behind the mirror, making it a virtual image, and it will be upright compared to the object.

A RAY DIAGRAM FOR A CONVEX MIRROR

What happens with a convex mirror? In this case the ray diagram looks like this:

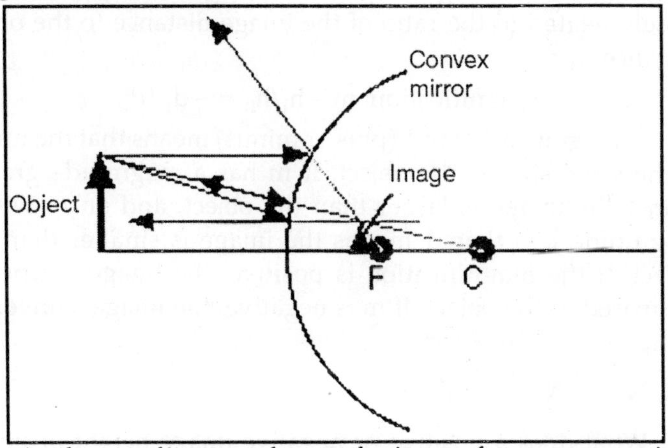

As the ray diagram shows, the image for a convex mirror is virtual, and upright compared to the object. A convex mirror is the kind of mirror used for security in stores, and is also the kind of mirror used on the passenger side of many cars ("Objects in mirror are closer than they appear."). A convex mirror will reflect a set of parallel rays in all directions; conversely, it will also take light from all directions and reflect it in one direction, which is exactly how it's used in stores and cars.

THE MIRROR EQUATION

Drawing a ray diagram is a great way to get a rough idea of how big the image of an object is, and where the image is located. We can also calculate these things precisely, using something known as the mirror equation.

The mirror equation: $1/d_0 + 1/d_i = 1/f$

where d_0 is the distance from the mirror to the objects, d_i is the distance from the mirror to the image, and f is the focal length of the mirror.

MAGNIFICATION

In most cases the height of the image differs from the height of the object, meaning that the mirror has done some magnifying (or reducing). The magnification, m, is defined as the ratio of the image height to the object height, which is closely related to the ratio of the image distance to the object distance:

$$\text{magnification: } m = h_i/h_0 = -d_i/d_0$$

A magnification of 1 (plus or minus) means that the image is the same size as the object. If m has a magnitude greater than 1 the image is larger than the object, and an m with a magnitude less than 1 means the image is smaller than the object. If the magnification is positive, the image is upright compared to the object; if m is negative, the image is inverted compared to the object.

SIGN CONVENTIONS

What does a positive or negative image height or image distance mean? To figure out what the signs mean, take the side of the mirror where the object is to be the positive side. Any distances measured on that side are positive. Distances measured on the other side are negative. f, the focal length, is positive for a concave mirror, and negative for a convex mirror. When the image distance is positive, the image is on the same side of the mirror as the object, and it is real and inverted. When the image distance is negative, the image is behind the mirror, so the image is virtual and upright.

A negative m means that the image is inverted. Positive means an upright image.

STEPS FOR ANALYZING MIRROR PROBLEMS

There are basically three steps to follow to analyse any mirror problem, which generally means determining where the image of an object is located, and determining what kind of image it is (real or virtual, upright or inverted).

- Step 1 - Draw a ray diagram. The more careful we are in constructing this, the better idea we 'll have of where the image is.
- Step 2 - Apply the mirror equation to determine the image distance. (Or to find the object distance, or the focal length, depending on what is given.)
- Step 3 - Make sure steps 1 and 2 are consistent with each other.

An Example

A Star Wars action figure, 8.0 cm tall, is placed 23.0 cm in front of a concave mirror with a focal length of 10.0 cm. Where is the image? How tall is the image? What are the characteristics of the image?

The first step is to draw the ray diagram, which should tell we that the image is real, inverted, smaller than the object, and between the focal point and the centre of curvature. The location of the image can be found from the mirror equation:

$$1/d_0 + 1/d_i = 1/f$$

$$\text{So } 1/d_i = 1/f - 11\text{-}1/d_0$$

which can be rearranged to:

$$d_i = d_0 f/(d_0 - f) = (23)(10)/(13) = 17.7\text{cm}$$

The image distance is positive, meaning that it is on the same side of the mirror as the object. This agrees with the ray diagram. Note that we don't need to worry about converting distances to meters; just make sure everything has the same units, and whatever unit goes into the equation is what comes out.

Calculating the magnification gives:

$$m = h_i/h_0 = -\, d_i/d_0 = -\, 17.7/23.0 = -0.77$$

Solving for the image height gives:

$$h_i = m\, h_0 = -\, 0.77 \times 8.0 = -\, 6.2 \text{ cm}$$

The negative sign for the magnification, and the image height, tells us that the image is inverted compared to the object.

To summarize, the image is real, inverted, 6.2 cm high, and 17.7 cm in front of the mirror.

Example - a Convex Mirror

The same Star Wars action figure, 8.0 cm tall, is placed 6.0 cm in front of a convex mirror with a focal length of -12.0 cm. Where is the image in this case, and what are the image characteristics?

Again, the first step is to draw a ray diagram. This should tell us that the image is located behind the mirror; that it is an upright, virtual image; that it is a little smaller than the object; and that the image is between the mirror and the focal point.

The second step is to confirm all those observations. The mirror equation, rearranged as in the first example, gives:

$$d_i = d_0 f/(d_0 - f) = (6)(-12)/(18) = -4.0 \text{ cm}$$

Solving for the magnification gives:

$$m = -d_i/d_0 = -(-4.0)/6.0 = 0.667$$

This gives an image height of 0.667 x 8 = 5.3 cm.

All of these results are consistent with the conclusions drawn from the ray diagram. The image is 5.3 cm high, virtual, upright compared to the object, and 4.0 cm behind the mirror.

Refraction

When we talk about the speed of light, we're usually talking about the speed of light in a vacuum, which is 3.00 x 10^8 m/s. When light travels through something else, such as glass, diamond, or plastic, it travels at a different speed. The speed of light in a given material is related to a quantity called the index of refraction, n, which is defined as the ratio of the speed of light in vacuum to the speed of light in the medium:

index of refraction : $n = c / v$

When light travels from one medium to another, the speed changes, as does the wavelength. The index of refraction can also be stated in terms of wavelength:

$$n = \lambda/\lambda_m$$

where λ is the wavelenght in vacuum and lm is the wavelenght in the medium.

Although the speed changes and wavelength changes, the frequency of the light will be constant. The frequency, wavelength, and speed are related by:

$$v = f\lambda$$

The change in speed that occurs when light passes from one medium to another is responsible for the bending of light, or refraction, that takes place at an interface. If light is travelling from medium 1 into medium 2, and angles are measured from the normal to the interface, the angle of transmission of the light into the second medium is related to the angle of incidence by Snell's law :

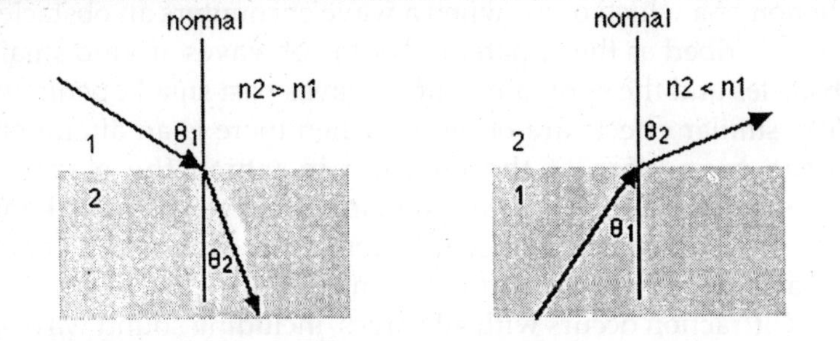

Snell's law : $n_1 \sin\theta_1 = n_2 \sin\theta_2$ or, equivalently, $\sin\theta_1 / \sin\theta_2 = v_1 / v_2$

Chapter 3

Diffraction

Diffraction is normally taken to refer to various phenomena which occur when a wave encounters an obstacle. It is described as the apparent bending of waves around small obstacles and the spreading out of waves past small openings. Very similar effects are observed when there is an alteration in the properties of the medium in which the wave is travelling, for example a variation in refractive index for light waves or in acoustic impedance for sound waves and these can also be referred to as diffraction effects.

Diffraction occurs with all waves, including sound waves, water waves, and electromagnetic waves such as visible light, x-rays and radio waves.

As physical objects have wave-like properties, diffraction also occurs with matter and can be studied according to the principles of quantum mechanics.

While diffraction occurs whenever propagating waves encounter such changes, its effects are generally most pronounced for waves where the wavelength is on the order of the size of the diffracting objects.

The complex patterns resulting from the intensity of a diffracted wave are a result of the superposition, or interference, of different parts of a wave that traveled to the observer by different paths.

The formalism of diffraction can also describe the way in which waves of finite extent propagate in free space. For example, the expanding profile of a laser beam, the beam shape of a radar antenna and the field of view of an ultrasonic transducer are all explained by diffraction theory.

EXAMPLES OF DIFFRACTION IN EVERYDAY LIFE

The effects of diffraction can be readily seen in everyday life. The most colorful examples of diffraction are those involving light; for example, the closely spaced tracks on a CD or DVD act as a diffraction grating to form the familiar rainbow pattern we see when looking at a disk. This principle can be extended to engineer a grating with a structure such that it will produce any diffraction pattern desired; the hologram on a credit card is an example.

Diffraction in the atmosphere by small particles can cause a bright ring to be visible around a bright light source like the sun or the moon. A shadow of a solid object, using light from a compact source, shows small fringes near its edges. The speckle pattern which is observed when laser light falls on an optically rough surface is also a diffraction phenomenon. All these effects are a consequence of the fact that light is a wave.

Diffraction can occur with any kind of wave. Ocean waves diffract around jetties and other obstacles. Sound waves can diffract around objects, this is the reason we can still hear someone calling us even if we are hiding behind a tree. Diffraction can also be a concern in some technical applications; it sets a fundamental limit to the resolution of a camera, telescope, or microscope.

THE EFFECTS OF DIFFRACTION OF LIGHT

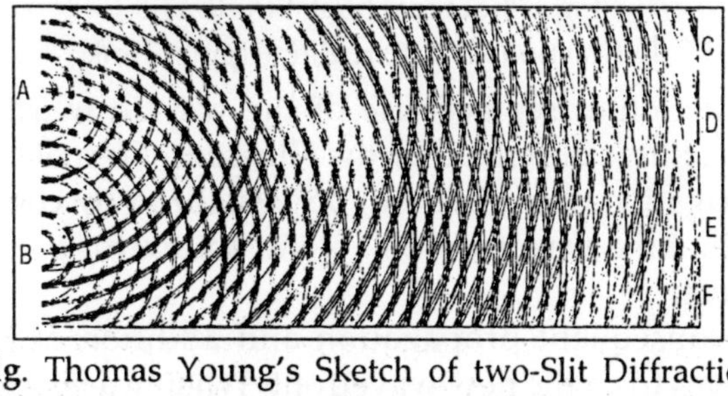

Fig. Thomas Young's Sketch of two-Slit Diffraction

The effects of diffraction of light were first carefully observed and characterized by Francesco Maria Grimaldi, who also coined the term *diffraction,* from the Latin *diffringere,* 'to break into pieces', referring to light breaking up into different directions. The results of Grimaldi's observations were published posthumously in 1665. Isaac Newton studied these effects and attributed them to *inflexion* of light rays. James Gregory observed the diffraction patterns caused by a bird feather, which was effectively the first diffraction grating.

In 1803 Thomas Young did his famous experiment observing interference from two closely spaced slits. Explaining his results by interference of the waves emanating from the two different slits, he deduced that light must propagate as waves.

Augustin-Jean Fresnel did more definitive studies and calculations of diffraction, published in 1815 and 1818, and thereby gave great support to the wave theory of light that had been advanced by Christiaan Huygens and reinvigorated by Young, against Newton's particle theory.

THE MECHANISM OF DIFFRACTION

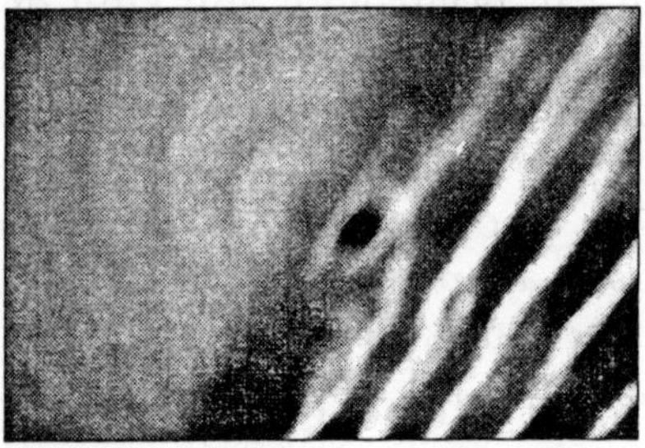

Fig. Photograph of Single-Slit Diffraction in a Circular Ripple Tank

Diffraction arises because of the way in which waves propagate; this is described by the Huygens–Fresnel principle. The propagation of a wave can be visualized by considering every point on a wavefront as a point source for a secondary radial wave.

The subsequent propagation and addition of all these radial waves form the new wavefront. When waves are added together, their sum is determined by the relative phases as well as the amplitudes of the individual waves, an effect which is often known as wave interference. The summed amplitude of the waves can have any value between zero and the sum of the individual amplitudes. Hence, diffraction patterns usually have a series of maxima and minima.

To determine the form of a diffraction pattern, we must determine the phase and amplitude of each of the Huygens wavelets at each point in space and then find the sum of these waves.

There are various analytical models which can be used to do this including the Fraunhoffer diffraction equation for the far field and the Fresnel Diffraction equation for the near-field. Most configurations cannot be solved analytically; solutions can be found using various numerical analytical methods including Finite element and boundary element methods

DIFFRACTION SYSTEMS

It is possible to obtain a qualitative understanding of many diffraction phenomena by considering how the relative phases of the individual secondary wave sources vary, and in particular, the conditions in which the phase difference equals half a cycle in which case waves will cancel one another out.

The simplest descriptions of diffraction are those in which the situation can be reduced to a two dimensional problem. For water waves, this is already the case, water waves propagate only on the surface of the water.

For light, we can often neglect one direction if the diffracting object extends in that direction over a distance far greater than the wavelength.

In the case of light shining through small circular holes we will have to take into account the full three dimensional nature of the problem. Some of the simpler cases of diffraction are considered below.

Single-Slit Diffraction

A long slit of infinitesimal width which is illuminated by light diffracts the light into a series of circular waves and the wavefront which emerges from the slit is a cylindrical wave of uniform intensity.

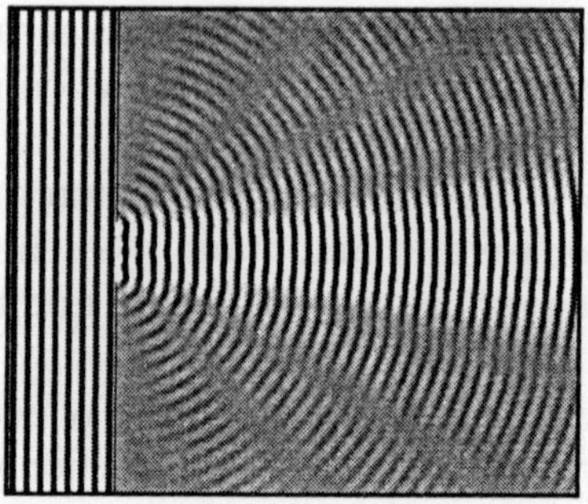

Fig. Numerical Approximation of Diffraction Pattern from a Slit of Width four Wavelengths with an Incident Plane Wave. The Main Central Beam, Nulls, and Phase Reversals are Apparent.

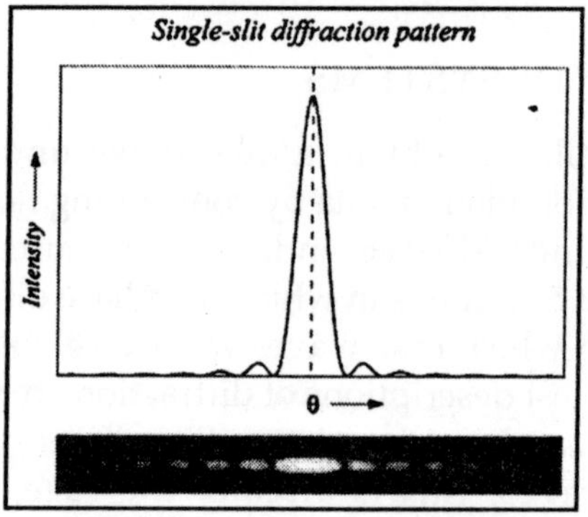

Fig. Graph and Image of Single-Slit Diffraction

A slit which is wider than a wavelength has a large number of point sources spaced evenly across the width of the slit. The light at a given angle is made up of contributions from each of these point sources and if the relative phases of these contributions vary by more than 2π, we expect to find minima and maxima in the diffracted light.

We can find the angle at which a first minimum is obtained in the diffracted light by the following reasoning. The light from a source located at the top edge of the slit interferes destructively with a source located at the middle of the slit, when the path difference between them is equal to $\lambda/2$.

Similarly, the source just below the top of the slit will

interfere destructively with the source located just below the middle of the slit at the same angle.

We can continue this reasoning along the entire height of the slit to conclude that the condition for destructive interference for the entire slit is the same as the condition for destructive interference between two narrow slits a distance apart that is half the width of the slit. The path difference is given by $\frac{d\sin(\theta)}{2}$ so that the minimum intensity occurs at an angle θ_{min} given by

$$d\sin\theta_{min} = \lambda$$

where d is the width of the slit.

A similar argument can be used to show that if we imagine the slit to be divided into four, six eight parts, etc, minima are obtained at angles θ_n given by

$$d\sin\theta_n = n\lambda$$

where n is an integer greater than zero.

There is no such simple argument to enable us to find the maxima of the diffraction pattern. The intensity profile can be calculated using the Fraunhofer diffraction integral as

$$I(\theta) = I_0 \operatorname{sinc}^2(d\sin\theta/\lambda)$$

where the sinc function is given by $\operatorname{sinc}(x) = \sin(\pi x)/(\pi x)$ if $x \neq 0$, and $\operatorname{sinc}(0) = 1$.

It should be noted that this analysis applies only to the far field, that is at a distance much larger than the width of the slit.

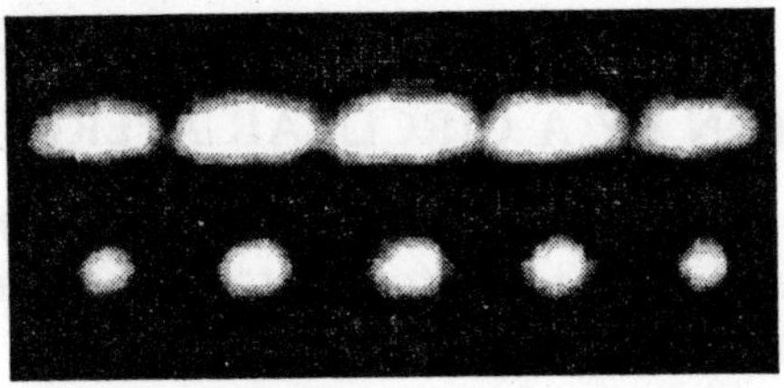

Fig. 2-Slit and 5-Slit Diffraction of Red Laser Light

DIFFRACTION GRATING

A diffraction grating is an optical component with a

regular pattern. The form of the light diffracted by a grating depends on the structure of the elements and the number of elements present, but all gratings have intensity maxima at angles θ_m which are given by the grating equation

$$d(\sin\theta_m + \sin\theta_i) = m\lambda.$$

where θ_i is the angle at which the light is incident, d is the separation of grating elements and m is an integer which can be positive or negative.

The light diffracted by a grating is found by summing the light diffracted from each of the elements, and is essentially a convolution of diffraction and interference patterns.

The figure shows the light diffracted by 2-element and 5-element gratings where the grating spacings are the same; it can be seen that the maxima are in the same position, but the detailed structures of the intensities are different.

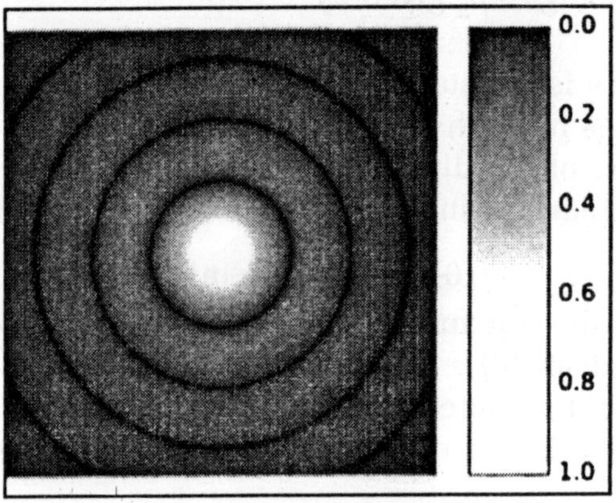

Fig. A computer-Generated Image of an Airy Disk

DIFFRACTION BY A CIRCULAR APERTURE

The far-field diffraction of a plane wave incident on a circular aperture is often referred to as the Airy Disk. The variation in intensity with angle is given by

$$I(\theta) = I_0\left(\frac{2J_1(ka\sin\theta)}{ka\sin\theta}\right)^2$$

where a is the radius of the circular aperture, k is equal to $2\pi/\lambda$ and J_1 is a Bessel function. The smaller the aperture, the larger the spot size at a given distance, and the greater the divergence of the diffracted beams.

PROPAGATION OF A LASER BEAM

The way in which the profile of a laser beam changes as it propagates is determined by diffraction. The output mirror of the laser is an aperture, and the subsequent beam shape is determined by that aperture.

Hence, the smaller the output beam, the quicker it diverges. Diode lasers have much greater divergence than He-Ne lasers for this reason.

Paradoxically, it is possible to reduce the divergence of a laser beam by first expanding it with one convex lens, and then collimating it with a second convex lens whose focal point is coincident with that of the first lens. The resulting beam has a larger aperture, and hence a lower divergence.

DIFFRACTION-LIMITED IMAGING

The ability of an imaging system to resolve detail is ultimately limited by diffraction. This is because a plane wave incident on a circular lens or mirror is diffracted. The light is not focused to a point but forms an Airy disk having a central spot in the focal plane with radius to first null of

$$d = 1.22\lambda N,$$

where λ is the wavelength of the light and N is the f-number (focal length divided by diameter) of the imaging optics. In object space, the corresponding angular resolution is

$$\sin\theta = 1.22\frac{\lambda}{D},$$

where D is the diameter of the entrance pupil of the imaging lens (e.g., of a telescope's main mirror).

As the point sources move closer together, the patterns will start to overlap, and ultimately they will merge to form a single pattern, in which case the two point sources cannot be resolved in the image. The Rayleigh criterion specifies that two point sources can be considered to be resolvable if the separation of the two images is at least the radius of the Airy disk, i.e. if the first minimum of one coincides with the maximum of the other. Thus, the larger the aperture of the lens, and the smaller the wavelength, the finer the resolution of an imaging system.

This is why telescopes have very large lenses or mirrors, and why optical microscopes are limited in the detail which they can see.

SPECKLE PATTERNS

The speckle pattern which is seen when using a laser pointer is another diffraction phenomenon. It is a result of the superpostion of many waves with different phases, which are produced when a laser beam illuminates a rough surface. They add together to give a resultant wave whose amplitude, and therefore intensity varies randomly.

COMMON FEATURES OF DIFFRACTION PATTERNS

Several qualitative observations can be made of diffraction in general:

- The angular spacing of the features in the diffraction pattern is inversely proportional to the dimensions of the object causing the diffraction, in other words: the smaller the diffracting object the 'wider' the resulting diffraction pattern and vice versa. (More precisely, this is true of the sines of the angles.)
- The diffraction angles are invariant under scaling; that is, they depend only on the ratio of the wavelength to the size of the diffracting object.
- When the diffracting object has a periodic structure, for example in a diffraction grating, the features generally become sharper. The third figure, for example, shows a comparison of a double-slit pattern with a pattern formed by five slits, both sets of slits having the same spacing, between the centre of one slit and the next.

PARTICLE DIFFRACTION

Quantum theory tells us that every particle exhibits wave properties. In particular, massive particles can interfere and therefore diffract. Diffraction of electrons and neutrons stood as one of the powerful arguments in favour of quantum mechanics. The wavelength associated with a particle is the de Broglie wavelength

$$\lambda = \frac{h}{p}$$

where h is Planck's constant and p is the momentum of the particle (mass × velocity for slow-moving particles). For most macroscopic objects, this wavelength is so short that it is not meaningful to assign a wavelength to them. A sodium atom traveling at about 3000 m/s would have a De Broglie wavelength of about 5 pico meters.

Because the wavelength for even the smallest of macroscopic objects is extremely small, diffraction of matter waves is only visible for small particles, like electrons, neutrons, atoms and small molecules. The short wavelength of these matter waves makes them ideally suited to study the atomic crystal structure of solids and large molecules like proteins. Relatively recently, larger molecules like buckyballs were also shown to diffract.

BRAGG DIFFRACTION

Diffraction from a three dimensional periodic structure such as atoms in a crystal is called Bragg diffraction. It is similar to what occurs when waves are scattered from a diffraction grating.

Bragg diffraction is a consequence of interference between waves reflecting from different crystal planes. The condition of constructive interference is given by *Bragg's law*:

$$m\lambda = 2d \sin\theta$$

where

λ is the wavelength,

d is the distance between crystal planes,

θ is the angle of the diffracted wave.

and *m* is an integer known as the *order* of the diffracted beam.

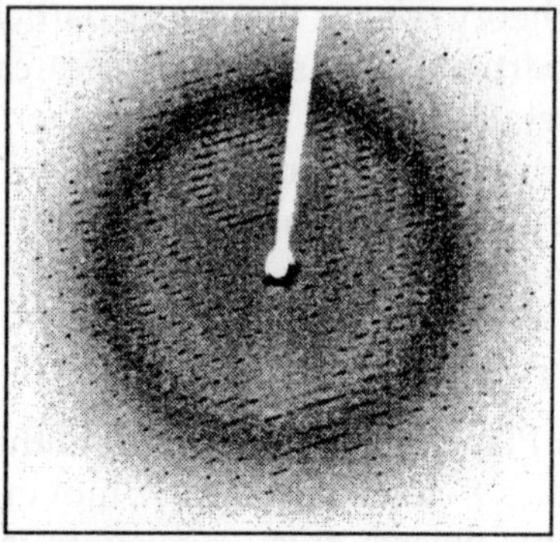

Fig. Following Bragg's Law, Each Dot (or *reflection*), in this Diffraction Pattern forms from the Constructive Interference of X-Rays Passing through a Crystal. The data can be used to Determine the Crystal's Atomic Structure.

Bragg diffraction may be carried out using either light of very short wavelength like x-rays or matter waves like neutrons (and electrons) whose wavelength is on the order of (or much smaller than) the atomic spacing. The pattern produced gives information of the separations of crystallographic planes *d*, allowing one to deduce the crystal structure. Diffraction contrast, in electron microscopes and x-topography devices in particular, is also a powerful tool for examining individual defects and local strain fields in crystals.

COHERENCE

The description of diffraction relies on the interference of waves emanating from the same source taking different paths to the same point on a screen. In this description, the difference in phase between waves that took different paths is only dependent on the effective path length.

This does not take into account the fact that waves that arrive at the screen at the same time were emitted by the source at different times. The initial phase with which the source emits waves can change over time in an unpredictable way. This means that waves emitted by the source at times that are too far apart can no longer form a constant interference pattern since the relation between their phases is no longer time independent.

The length over which the phase in a beam of light is correlated, is called the Coherence length. In order for interference to occur, the path length difference must be smaller than the coherence length.

This is sometimes referred to as spectral coherence as it is related to the presence of different frequency components in the wave. In the case light emitted by an atomic transition, the coherence length is related to the lifetime of the excited state from which the atom made its transition.

If waves are emitted from an extended source this can lead to incoherence in the transversal direction. When looking at a cross section of a beam of light, the length over which the phase is correlated is called the transverse coherence length.

In the case of Young's double slit experiment this would mean that if the transverse coherence length is smaller than the spacing between the two slits the resulting pattern on a screen would look like two single slit diffraction patterns.

In the case of particles like electrons, neutrons and atoms, the coherence length is related to the spacial extent of the wave function that describes the particle.

Chapter 4

Interference

When two or more waves interact and combine, they interfere with one another. But interference is not necessarily bad: waves may interfere constructively, resulting in a wave larger than the original waves. Or, they may interfere destructively, combining in such a way that they form a wave smaller than the original ones.

Even so, destructive interference may have positive effects: without the application of destructive interference to the muffler on an automobile exhaust system, for instance, noise pollution from cars would be far worse than it is. Other examples of interference, both constructive and destructive, can be found wherever there are waves: in water, in sound, in light.

WAVES

Whenever energy ripples through space, there is a wave. In fact, wave motion can be defined as a type of harmonic motion (repeated movement of a particle about a position of equilibrium, or balance) that carries energy from one place to another without actually moving any matter. A wave on the ocean is an example of a mechanical wave, or one that involves matter; but though the matter moves in place, it is only the energy in the wave that experiences net movement.

Wave motion is related to oscillation, a type of harmonic motion in one or more dimensions. There is one critical difference, however: oscillation involves no net movement, only movement in place, whereas the harmonic motion of waves carries energy from one place to another. Yet, individual waves themselves are oscillating even as the overall wave pattern moves.

A transverse wave forms a regular up-and-down pattern in which the oscillation is perpendicular to the direction the wave is moving. Ocean waves are transverse, though they also have properties of longitudinal waves. In a longitudinal wave, of which a sound wave is the best example, oscillation occurs in the same direction as the wave itself.

Parameters of Wave Motion

Some waves, composed of pulses, do not follow regular patterns. However, the waves of principal concern in the present context are periodic waves, ones in which a uniform series of crests and troughs follow each other in regular succession. Periodic motion is movement repeated at regular intervals called periods. In the case of wave motion, a period (represented by the symbol *T*) is the amount of time required to complete one full cycle of the wave, from trough to crest and back to trough.

Period can be mathematically related to several other aspects of wave motion, including wave speed, frequency, and wavelength. Frequency (abbreviated *f*) is the number of waves passing through a given point during the interval of one second. It is measured in Hertz (Hz), named after nineteenth-century German physicist Heinrich Rudolf Hertz, and a Hertz is equal to one cycle of oscillation per second.

Higher frequencies are expressed in terms of kilohertz (kHz; 10^3 or 1,000 cycles per second) or megahertz (MHz; 10^6 or 1 million cycles per second.) Wavelength (represented by the symbol abbreviated λ, the Greek letter lambda) is the distance between a crest and the adjacent crest, or a trough and an adjacent trough, of a wave. The higher the frequency, the shorter the wavelength.

Another parameter for describing wave motion—one that is mathematically independent from the quantities so far described—is amplitude, or the maximum displacement of particles from a position of stable equilibrium. For an ocean wave, amplitude is the distance from either the crest or the trough to the level that the ocean would maintain if it were perfectly still.

SUPERPOSITION AND INTERFERENCE

Superposition

The principle of superposition holds that when several individual but similar physical events occur in close proximity, the resulting effect is the sum of the magnitude of the separate events. This is akin to the popular expression, "The whole is greater than the sum of the parts," and it has numerous applications in physics.

Where the strength of a gravitational field is being measured, for instance, superposition dictates that the strength of that field at any given point is the sum of the mass of the individual particles in that field. In the realm of electromagnetic force, the same statement applies, though the units being added are electrical charges or magnetic poles, rather than quantities of mass. Likewise, in an electrical circuit, the total current or voltage is the sum of the individual currents and voltages in that circuit.

Superposition applies only in equations for linear events—that is, phenomena that involve movement along a straight line. Waves are linear phenomena, and, thus, the principle describes the behaviour of all waves when they come into contact with one another. If two or more waves enter the same region of space at the same time, then, at any instant, the total disturbance produced by the waves at any point is equal to the sum of the disturbances produced by the individual waves.

Interference

The principle of superposition does not require that waves actually combine; rather, the net effect is as though they were combined. The actual combination or joining of two or more waves at a given point in space is called interference, and, as a result, the waves produce a single wave whose properties are determined by the properties of the individual waves.

If two waves of the same wavelength occupy the same space in such a way that their crests and troughs align, the wave they produce will have an amplitude greater than that possessed by either wave initially. This is known as constructive interference. The more closely the waves are in phase—that is, perfectly aligned—the more constructive the interference.

It is also possible that two or more waves can come together such that the trough of one meets the crest of the other, or vice versa. In this case, what happens is destructive interference, and the resulting amplitude is the difference between the values for the individual waves. If the waves are perfectly unaligned—in other words, if the trough of one exactly meets the crest of the other—their amplitudes cancel out, and the result is no wave at all.

Resonance

It is easy to confuse interference with resonance, and, therefore, a word should be said about the latter phenomenon. The term resonance describes a situation in which force is applied to an oscillator at the point of maximum amplitude. In this way, the motion of the outside force is perfectly matched to that of the oscillator, making possible a transfer of energy. As with interference, resonance implies alignment between two physical entities; however, there are several important differences.

Resonance can involve waves, as, for instance, when sound waves resonate with the vibrations of an oscillator, causing a transfer of energy that sometimes produces dramatic results. But in these cases, a wave is interacting with an oscillator, not a wave with a wave, as in situations of interference. Furthermore, whereas resonance entails a transfer of energy, interference involves a combination of energy.

Transfer Vs. Combination

The importance of this distinction is easy to see if one substitutes money for energy, and people for objects. If one passes on a sum of money to another person, a business, or an institution—as a loan, repayment of a loan, a purchase, or a gift—this is an example of a transfer. On the other hand, when married spouses each earn paychecks, their cash is combined.

Transfer thus indicates that the original holder of the cash (or energy) no longer has it. Yet, if the holder of the cash combines funds with those of another, both share rights to an amount of money greater than the amount each originally owned. This is analogous to constructive interference.

On the other hand, a husband and wife (or any other group of people who pool their cash) also share liabilities, and, thus, a married person may be subject to debt incurred by his or her spouse. If one spouse creates debt so great that the other spouse cannot earn enough to maintain the payments, this painful situation is analogous to destructive interference.

REAL-LIFE APPLICATIONS

Mechanical Waves

One of the easiest ways to observe interference is by watching the behaviour of mechanical waves. Drop a stone into a still pond, and watch how its waves ripple: this, as with most waveforms in water, is an example of a surface wave, or one that displays aspects of both transverse and longitudinal wave motion. Thus, as the concentric circles of a longitudinal wave ripple outward in one dimension, there are also transverse movements along a plane perpendicular to that of the longitudinal wave.

While the first wave is still rippling across the water, drop another stone close to the place where the first one was dropped. Now, there are two surface waves, crests and troughs colliding and interfering. In some places, they will interfere constructively, producing a wave—or rather, a portion of a wave—that is greater in amplitude than either of the original waves. At other places, there will be destructive interference, with some waves so perfectly out of phase that at one instant in time, a given spot on the water may look as though it had not been disturbed at all.

One of the interesting aspects of this interaction is the lack of uniformity in the instances of interference. As suggested in the preceding paragraph, it is usually not entire waves, but merely portions of waves, that interfere constructively or destructively. The result is that a seemingly simple event—dropping two stones into a still pond—produces a dazzling array of colliding circles, broken by outwardly undisturbed areas of destructive interference.

A similar phenomenon, though manifested by the interaction of geometric lines rather than concentric circles, occurs when two power boats pass each other on a lake. The first boat chops up the water, creating a wake that widens behind it: when seen from the air, the boat appears to be at the apex of a triangle whose sides are formed by rippling eddies of water.

Now, another boat passes through the wake of the first, only it is going in the opposite direction and producing its own ever-widening wake as it goes. As the waves from the two boats meet, some are in phase, but, more often than not, they are only partly in phase, or they possess differing wavelengths. Therefore, the waves at least partially cancel out one another in places, and in other places, reinforce one another. The result is an interesting patchwork of patterns when seen from the air.

SOUND WAVES

In Tune and Out of Tune

The relationships between musical notes can be intriguing, and though tastes in music vary, most people know when music is harmonious and when it is discordant. As discussed in the essay on frequency, this harmony or discord can be equated to the mathematical relationships between the frequencies of specific notes: the lower the numbers involved in the ratio, the more pleasing the sound.

The ratio between the frequency of middle C and that of

its first harmonic—that is, the C note exactly one octave above it—is a nice, clean 1:2. If one were to play a song in the key of C-which, on a piano, involves only the "white notes" C-D-E-F-G-A-B—everything should be perfectly harmonious and (presumably) pleasant to the ear. But what if the piano itself is out of tune? Or what if one key is out of tune with the others?

The result, for anyone who is not tone-deaf, produces an overall impression of unpleasantness: it might be a bit hard to identify the source of this discomfort, but it is clear that something is amiss. At best, an out-of-tune piano might sound like something that belonged in a saloon from an old Western; at worst, the sound of notes that do not match their accustomed frequencies can be positively grating.

Tuning Fork

To rectify the situation, a professional piano tuner uses a tuning fork, an instrument that produces a single frequency—say, 264 Hz, which is the frequency of middle C. The piano tuner strikes the tuning fork, and at the same time strikes the appropriate key on the piano. If their frequencies are perfectly aligned, so is the sound of both; but, more likely, there will be interference, both constructive and destructive.

As time passes—measured in seconds or even fractions of seconds—the sounds of the tuning fork and that of the piano key will alternate between constructive and destructive interference. In the case of constructive interference, their combined sound will become louder than the individual sounds of either; and when the interference is destructive, the sound of both together will be softer than that produced by either the fork or the key.

The piano tuner listens for these fluctuations of loudness, which are called beats, and adjusts the tension in the appropriate piano string until the beats disappear completely. As long as there are beats, the piano string and the tuning fork will produce together a frequency that is the average of the two: if, for instance, the out-of-tune middle C string vibrates at 262 Hz, the resulting frequency will be 263 Hz.

Difference Tones

Another interesting aspect of the interaction between notes is the "difference tone," created by discord, which the human ear perceives as a third tone. Though E and F are both part of the C scale, when struck together, the sound is highly discordant. In light of what was said about ratios between frequencies, this dissonance is fitting, as the ratio here involves relatively high numbers—15:16.

When two notes are struck together, they produce a combination tone, perceived by the human ear as a third tone. If the two notes are harmonious, the "third tone" is known as a summation tone, and is equal to the combined frequencies of the two notes. But if the combination is dissonant, as in the case of E and F, the third tone is known as a difference tone, equal to the difference in frequencies. Since an E note vibrates at 330 Hz, and an F note at 352 Hz, the resulting difference tone is equal to 22 Hz.

Destructive Interference in Sound Waves

When music is played in a concert hall, it reverberates off the walls of the auditorium. Assuming the place is well designed acoustically, these bouncing sound waves will interfere constructively, and the auditorium comes alive with the sound of the music. In other situations, however, the sound waves may interfere destructively, and the result is a certain muffled deadness to the sound.

Clearly, in a music hall, destructive interference is a problem; but there are cases in which it can be a benefit—situations, that is, in which the purpose, indeed, is to deaden the sound. One example is an automobile muffler. A car's exhaust system makes a great deal of noise, and, thus, if a car does not have a proper muffler, it creates a great deal of noise pollution. A muffler counteracts this by producing a sound wave out of phase with that of the exhaust system; hence, it cancels out most of the noise.

Destructive interference can also be used to reduce sound in a room. Once again, a machine is calibrated to generate sound waves that are perfectly out of phase with the offending noise—say, the hum of another machine. The resulting effect conveys the impression that there is no noise in the room, though, in fact, the sound waves are still there; they have merely canceled each other out.

Electromagnetic Waves

In 1801, English physicist Thomas Young, known for Young's modulus of elasticity became the first scientist to identify interference in light waves. Challenging the corpuscular theory of light put forward by Sir Isaac Newton, Young set up an experiment in which a beam of light passed through two closely spaced pinholes onto a screen. If light was truly made of particles, he said, the beams would project two distinct points onto the screen. Instead, what he saw was a pattern of interference.

In fact, Newton was partly right, but Young's discovery helped advance the view of light as a wave, which is also partly right. (According to quantum theory, developed in the twentieth century, light behaves both as waves and as particles.) The interference in the visible spectrum that Young witnessed was manifested as bright and dark bands. These bands are known as fringes—variations in intensity not unlike the beats created in some instances of sound interference.

Oily Films and Rainbows

Many people have noticed the strangely beautiful pattern of colors generated when light interacts with an oily substance, as when light reflected on a soap bubble produces an astonishing array of shades. Sometimes, this can happen in situations not otherwise aesthetically pleasing: an oily film in a parking lot, left there by a car's leaky crankcase, can produce a rainbow of colors if the sunlight hits it just right.

This happens because the thickness of the oil causes a delay in reflection of the light beam. Some colors pass through the film, becoming delayed and, thus, getting out of phase with the reflected light on the surface of the film. These shades destructively interfere to such an extent that the waves are cancelled, rendering them invisible. Other colors reflect off the surface so that they are perfectly in phase with the light traveling through the film, and appear as an attractive swirl of color on the surface of the oil.

The phenomenon of light-wave interference with oily or filmy surfaces has the effect of filtering light, and, thus, has a number of applications in areas relating to optics: sunglasses, lenses for binoculars or cameras, and even visors for astronauts. In each case, unfiltered light could be harmful or, at least, inconvenient for the user, and the destructive interference eliminates certain colors and unwanted reflections.

Radio Waves

Visible light is only a small part of the electromagnetic spectrum, whose broad range of wave phenomena are, likewise, subject to constructive or destructive interference. After visible light, the area of the spectrum most people experience during an average day is the realm of relatively low-frequency, long-wave length radio waves and microwaves, the latter including television broadcast signals.

People who rely on an antenna for their TV reception are likely to experience interference at some point. However, an increasing number of Americans use either cable or satellite systems to pick up TV programs. These are much less susceptible to interference, due to the technology of coaxial cable, on the one hand, and digital compression, on the other. Thus, interference in television reception is a gradually diminishing problem.

Sci-Tech Encyclopedia: Electrical interference

Disturbance to the normal or expected operation of electrical or electronic devices, equipment, and systems. Electrical interference is sometimes called radio-frequency interference (RFI) or electromagnetic interference (EMI). Electrical noise is a broader term that includes those phenomena that are generally termed electrical interference, but also includes naturally occurring currents or voltages that are more or less continuous and cannot be completely eliminated.

Electrical interference originates from one or more of the following: transmitters such as those used for broadcast, communication, radar, and navigation; artificial incidental emission sources such as from sparking of motor brushes, automotive ignition, and fluorescent lamps; and natural phenomena such as lightning and electrostatic discharges. The interference emissions may be radiated through space or conducted along the paths of electrical wires in power lines and signal cables.

There are several means of mitigating electrical interference, including grounding, bonding, shielding, filtering, and transient control. There are two major approaches. One is to reduce radiation from boxes, cabinets, racks, and consoles by the design of multilayer printed circuit boards and backplanes, or to use metal or metallized plastic cases with EMI-protected apertures. The other involves reduction of radiation to or from the interconnecting cable by using filter pin connectors or foil-wrap shields terminated with complete coverage by the connector backshell at the respective box bulkhead.

INTERFERENCE IN SOUND WAVES

When two sound waves occur at the same time and are in the same phase, i.e., when the condensations of the two coincide and hence their rarefactions also, the waves reinforce each other and the sound becomes louder. This is known as constructive interference. On the other hand, two sound waves occurring simultaneously and having the same intensity neutralize each other if the rarefactions of the one coincide with the condensations of the other, i.e., if they are of opposite phase. This canceling is known as destructive interference. In this case, the result is silence.

Alternate reinforcement and neutralization (or weakening) take place when two sound waves differing slightly in frequency are superimposed. The audible result is a series of pulsations or, as these pulsations are called commonly, beats, caused by the alternate coincidence of first a condensation of the one wave with a condensation of the other and then a condensation with a rarefaction. The beat frequency is equal to the difference between the frequencies of the interfering sound waves.

Interference in Light Waves

Light waves reinforce or neutralize each other in very much the same way as sound waves. If, for example, two light waves each of one color (monochromatic waves), of the same amplitude, and of the same frequency are combined, the interference they exhibit is characterized by so-called fringes—a series of light bands (resulting from reinforcement) alternating with dark bands (caused by neutralization).

Such a pattern is formed either by light passing through two narrow slits and being diffracted, or by light passing through a single slit. In the case of two slits, each slit acts as a light source, producing two sets of waves that may combine or cancel depending upon their phase relationship. In the case of a single slit, each point within the slit acts as a light source. In all cases, for light waves to demonstrate such behaviour, they must emanate from the same source; light from distinct sources has too many random differences to permit interference patterns. The relative positions of light and dark lines depend upon the wavelength of the light, among other factors. Thus, if white light, which is made up of all colors, is used instead of monochromatic light, bands of color are formed because each color, or wavelength, is reinforced at a different position. This fact is utilized in the diffraction grating, which forms a spectrum by diffraction and interference of a beam of light incident on it. Newton's rings also are the result of the interference of light.

They are formed concentrically around the point of contact between a glass plate and a slightly convex lens set upon it or between two lenses pressed together; they consist of bright rings separated by dark ones when monochromatic light is used, or of alternate spectrum-colored and black rings when white light is used. Various natural phenomena are the result of interference, e.g., the colors appearing in soap bubbles and the iridescence of mother-of-pearl and other substances.

Interference as a Scientific Tool

The experiments of Thomas Young first illustrated interference and definitely pointed the way to a wave theory of light. A. J. Fresnel's experiments clearly demonstrated that the interference phenomena could be explained adequately only upon the basis of a wave theory.

The thickness of a very thin film such as the soap-bubble wall can be measured by an instrument called the interferometer. When the wavelength of the light is known, the interferometer indicates the thickness of the film by the interference patterns it forms. The reverse process, i.e., the measurement of the length of an unknown light wave, can also be carried out by the interferometer.

The Michelson interferometer used in the Michelson-Morley experiment of 1887 to determine the velocity of light had a half-silvered mirror to split an incident beam of light into two parts at right angles to one another.

The two halves of the beam were then reflected off mirrors and rejoined. Any difference in the speed of light along the paths could be detected by the interference pattern. The failure of the experiment to detect any such difference threw doubt on the existence of the ether and thus paved the way for the special theory of relativity.

Another type of interferometer devised by Michelson has been applied in measuring the diameters of certain stars. The radio interferometer consists of two or more radio telesçopes separated by fairly large distances (necessary because radio waves are much longer than light waves) and is used to pinpoint and study various celestial sources of radiation in the radio range. Astronomical interferometers consisting of two or more optical telescopes are used to enhance visible images of distant celestial objects.

In the law of patents, the presence of two pending applications, or an existing patent and a pending application that encompass an identical invention or discovery.

When interference exists, the Patent and Trademark Office conducts an investigation to ascertain the priority of invention between the conflicting applications, or the application and the patent. A patent is customarily granted to the earlier invention.

INTERFERENCE OF TWO CIRCULAR WAVES

Fig. Interference of Two Circular Waves

Interference of two circular waves - Wavelength (decreasing bottom to top) and Wave centers distance (increasing to the right). Absolute value snapshots of the (real-valued, scalar) wave field. As time progresses, the wave fronts would move outwards from the two centers, but the dark regions (destructive interference) stay fixed.

In physics, interference is the addition (superposition) of two or more waves that result in a new wave pattern.

As most commonly used, the term interference usually refers to the interaction of waves which are correlated or coherent with each other, either because they come from the same source or because they have the same or nearly the same frequency.

Two non-monochromatic waves are only fully coherent with each other if they both have exactly the same range of wavelengths and the same phase differences at each of the constituent wavelengths.

The total phase difference is derived from the sum of both the path difference and the initial phase difference (if the waves are generated from 2 or more different sources). It can then be concluded whether the waves reaching a point are *in phase* (constructive interference) or *out of phase* (destructive interference).

The principle of superposition of waves states that the resultant displacement at a point is equal to the vector sum of the displacements of different waves at that point. If a crest of a wave meets a crest of another wave at the same point then the crests interfere *constructively* and the resultant wave amplitude is greater. If a crest of a wave meets a trough of another wave then they interfere *destructively*, and the overall amplitude is decreased.

This form of interference can occur whenever a wave can propagate from a source to a destination by two or more paths of different length. Two or more sources can only be used to produce interference when there is a fixed phase relation between them, but in this case the interference generated is the same as with a single source.

EXPERIMENTS

Thomas Young's double-slit experiment showed interference phenomena where two beams of light which are coherent interfere to produce a pattern.

The beams of light both have the same wavelength range and at the centre of the interference pattern. They have the same phases at each wavelength, as they both come from the same source.

INTERFERENCE PATTERNS

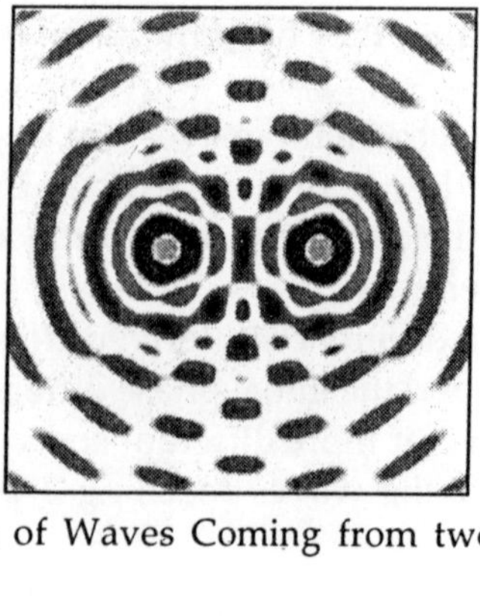

Fig. Interference of Waves Coming from two Point Sources.

For two coherent sources, the spatial separation between sources is half the wavelength times the number of nodal lines.

Light from any source can be used to obtain interference patterns, for example, Newton's rings can be produced with sunlight. However, in general white light is less suited for producing clear interference patterns, as it is a mix of a full spectrum of colours, that each have different spacing of the interference fringes. Sodium light is close to monochromatic and is thus more suitable for producing interference patterns. The most suitable is laser light because it is almost perfectly monochromatic.

CONSTRUCTIVE AND DESTRUCTIVE INTERFERENCE

Fig. Interference Pattern Produced with a Michelson Interferometer. Bright Bands are the Result of Constructive Interference while the Dark Bands are the Result of Destructive Interference.

Consider two waves that are in phase,with amplitudes A_1 and A_2. Their troughs and peaks line up and the resultant wave will have amplitude $A = A_1 + A_2$. This is known as constructive interference.

If the two waves are π radians, or 180°, out of phase, then one wave's crests will coincide with another wave's troughs and so will tend to cancel out. The resultant amplitude is $A =$ $|A_1$ " $A_2|$. If $A_1 = A_2$, the resultant amplitude will be zero. This is known as destructive interference.

When two sinusoidal waves superimpose, the resulting

waveform depends on the frequency (or wavelength) amplitude and relative phase of the two waves.

If the two waves have the same amplitude A and wavelength the resultant waveform will have an amplitude between 0 and $2A$ depending on whether the two waves are in phase or out of phase.

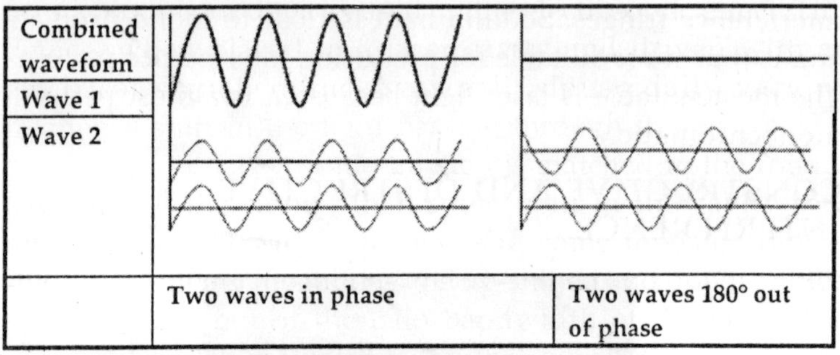

GENERAL QUANTUM INTERFERENCE

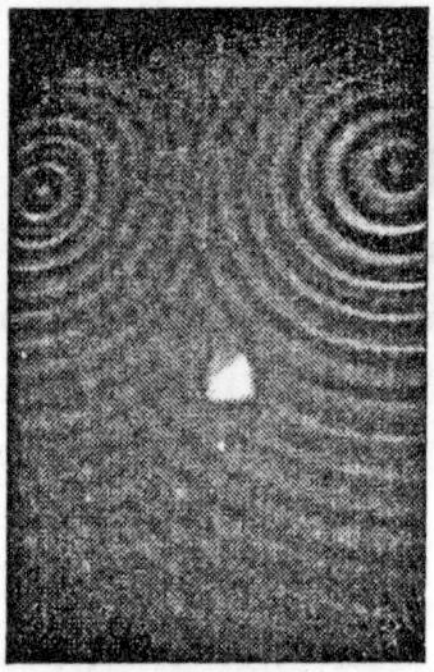

Fig. Two Point Interference In a Ripple Tank.

If a system is in state Ψ its wavefunction is described in Dirac or bra-ket notation as:

$$|\psi\rangle = \sum_i |i\rangle \psi_i$$

where $|i\rangle_s$ the s specify the different quantum "alternatives" available (technically, they form an eigenvector basis) and the Ψ_i are the probability amplitude coefficients, which are complex numbers.

The probability of observing the system making a transition or quantum leap from state Ψ to a new state Φ is the square of the modulus of the scalar or inner product of the two states:

$$prob(\psi \Rightarrow 0) = |\langle\psi|\phi\rangle|^2 = \left|\sum_i \psi_i^* \varphi_i\right|^2$$

$$= \sum_{ij} \psi_i^* \psi_j \psi_j^* \varphi_i = \sum_i |\psi_i|^2 |\varphi_i|^2 + \sum_{ij;i\neq j} \psi_i^* \psi_j \psi_j^* \varphi_i$$

where $\psi_i = \langle i|\psi\rangle$ and similarly $\phi_i = \langle i|\phi\rangle$ are the coefficients of the final state of the system. * is the complex conjugate so that $\varphi_i = \langle i|\phi\rangle$, etc.

Now let's consider the situation classically and imagine that the system transited from to via an intermediate state $|i\rangle$. Then we would *classically* expect the probability of the two-step transition to be the sum of all the possible intermediate steps. So we would have

$$prob(\psi \Rightarrow \phi) = \sum_{ij} prob(\psi \Rightarrow i \Rightarrow \varphi_i)$$

$$= \sum_i |\langle\psi|i\rangle|^2 |\langle i|\phi\rangle|^2$$

$$= \sum_i |\psi_i|^2 |\varphi_i|^2.$$

The classical and quantum derivations for the transition probability differ by the presence, in the quantum case, of the extra terms $\sum_{ij;i\neq j} \psi_i^* \psi_j \psi_j^* \varphi_i$; these extra quantum terms represent *interference* between the different $i \neq j$ intermediate "alternatives". These are consequently known as the quantum interference terms, or cross terms. This is a purely quantum effect and is a consequence of the non-additivity of the probabilities of quantum alternatives.

The interference terms vanish, via the mechanism of quantum decoherence, if the intermediate state $|i\rangle$ is measured or coupled with the environment.

EXAMPLES

A conceptually simple case of interference is a small (compared to wavelength) source - say, a small array of regularly spaced small sources.

Consider the case of a flat boundary (say, between two media with different densities or simply a flat mirror), onto which the plane wave is incident at some angle.

In this case of continuous distribution of sources, constructive interference will only be in specular direction - the direction at which angle with the normal is exactly the same as the angle of incidence. Thus, this results in the law of reflection which is simply the result of constructive interference of a plane wave on a plane surface.

Chapter 5

Two-Point Source Interference

SOURCE INTERFERENCE PATTERN

The nature of the interference pattern produced by two bobbing sources in a ripple tank (water tank) was discussed. The diagram at the right depicts the pattern resulting from the propagation of water waves across the surface of the water. The waves propagate outward from the point sources, forming a series of concentric circles about the source.

In the diagram, the thick lines represent wave crests and the thin lines represent wave troughs. The crests and troughs from the two sources interfere with each other at a regular rate to produce nodes (pictured in blue on the diagram) and antinodes (pictured in red) along the water surface.

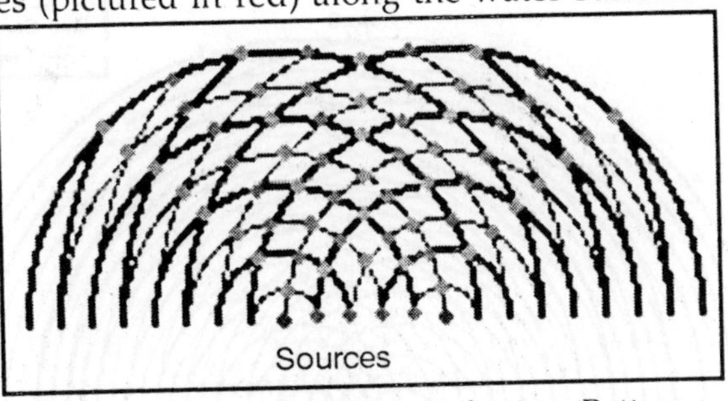

Fig. Two-Point Source Interference Pattern

The nodal positions are locations where the water is undisturbed; the antinodal positions are locations where the water is undergoing maximum disturbances above and below the surrounding water level. One unique feature of the two-point source interference pattern is that the antinodal and nodal positions all lie along distinct lines. Each line can be described as a relatively straight hyperbola.

The spatial separation between the antinodal and nodal lines in the pattern is related to the wavelength of the waves. The mathematical relationship will be explored later in this lesson. For now, we will investigate the underlying causes of this unique pattern and introduce some nomenclature (naming conventions) which will be utilized throughout the lesson. A series of concentric circles about two point sources (labeled as S_1 and S_2).

The pattern could be the result of water waves in a ripple tank resulting from two vibrating sources; or the result of sound waves from two speakers traveling through a room; or the result of two light waves moving through a room after passing through two slits or pinholes in a sheet of paper. The thick lines represent wave crests and the thin lines represent wave troughs.

The red dots in the represent the antinodal positions; the blue dots represent the nodal positions. The red lines drawn through the antinodal points (red dots) are referred to as antinodal lines and the blue lines drawn through the nodal points (blue dots) are referred to as nodal lines.

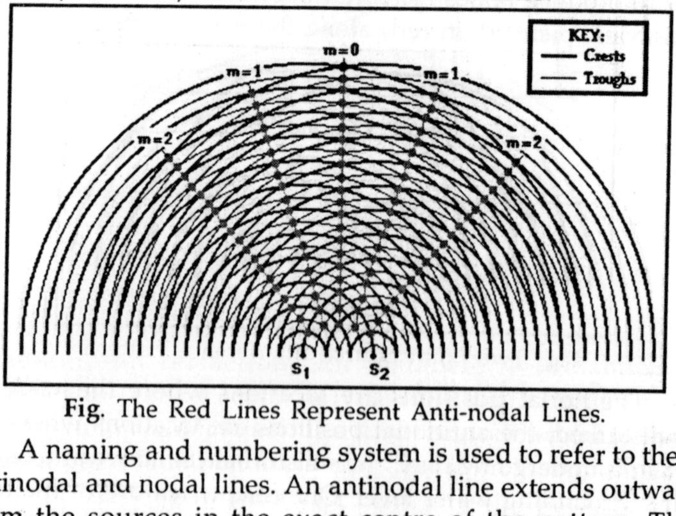

Fig. The Red Lines Represent Anti-nodal Lines.

A naming and numbering system is used to refer to these antinodal and nodal lines. An antinodal line extends outward from the sources in the exact centre of the pattern. This antinodal line is referred to as the central antinodal line. More antinodal lines are present to the left and to the right of the central antinodal line.

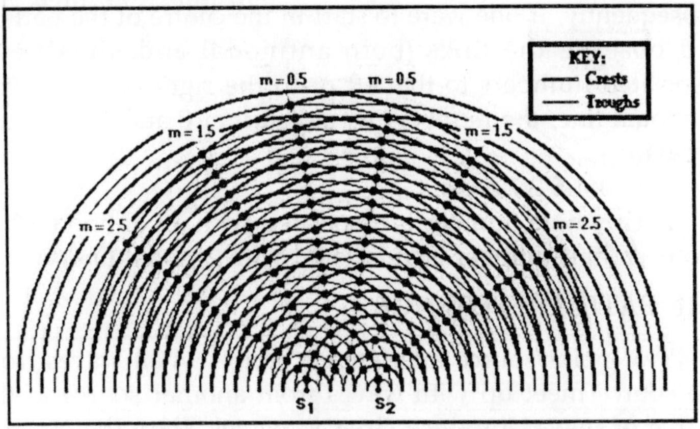

Fig. The Blue Lines Represent Nodal Lines.

These are referred to as the first antinodal line, the second antinodal line, the third antinodal line (if present), etc. Each antinodal line is separated by a nodal line. The nodal lines are also named; the first nodal line to the left or to the right of the central antinodal line is referred to as the first nodal line. The second nodal line and the third nodal line are found as one moves further to the left and to the right of the centre of the pattern.

Each line in the pattern is assigned a number, known as the order number and represented by the letter m. The numbering system associated with this pattern is just as creative as the naming system. The central antinodal line is assigned an order number of 0. The first antinodal line is assigned an order number of 1; the second antinodal line is assigned an order number of 2; the third antinodal line is assigned an order number of 3; etc. Nodal lines are assigned half-numbers. The first nodal line, located between the central antinodal line (m = 0) and the first antinodal line (m = 1) is assigned the order number of 0.5.

The second nodal line, located between the first antinodal line (m = 1) and the second antinodal line (m = 2) is assigned the order number of 1.5. Finally, the third nodal line, located between the second antinodal line (m = 2) and the third antinodal line (m = 3) is assigned the order number of 2.5. Subsequently, if one were to start in the centre of the pattern and observe the lines (both antinodal and nodal) and associated numbers to the left or to the right of the central antinodal line, the numbers would start at 0 and increase by one-half:

0 0.5 1 1.5 2 2.5 etc.

Each whole number is associated with an antinodal line and each half-number is associated with a nodal line.

THE PATH DIFFERENCE

Two-point source interference occurs when waves from one source meet up with waves from another source. If the source of waves produce circular waves, then the circular wavefronts will meet within the medium to produce a pattern. The pattern is characterized by a collection of nodes and antinodes which lie along nearly straight lines referred to as antinodal lines and nodal lines.

If the wave sources have identical frequencies, then there will be an antinodal line in the exact centre of the pattern and an alternating series of nodal and antinodal lines to the left and the right of the central antinodal line. Each line in the pattern is assigned a name (e.g., first antinodal line) and an order number (represented by the symbol m). A representative two-point source interference pattern with accompanying order numbers (m values) is shown below.

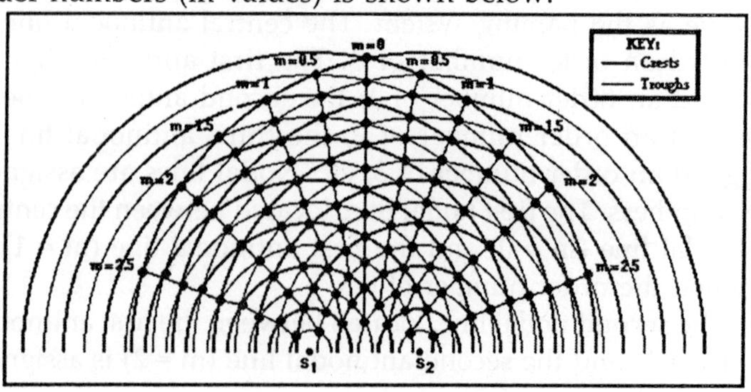

We will investigate the rationale behind the numbering system and develop some mathematical equations which relate the features of the pattern to the wavelength of the waves.

This investigation will involve the analysis of several antinodal and nodal locations on a typical two-point source interference pattern. It will be assumed in the discussion that the wave sources are producing waves with identical frequencies (and therefore identical wavelengths).

To begin, consider the pattern shown in the below. Point A is a point located on the first antinodal line. This specific antinode is formed as the result of the interference of a crest from Source 1 (S_1) meeting up with a crest from Source 2 (S_2). The two wave crests are taking two different paths to the same location to constructively interfere to form the antinodal point.

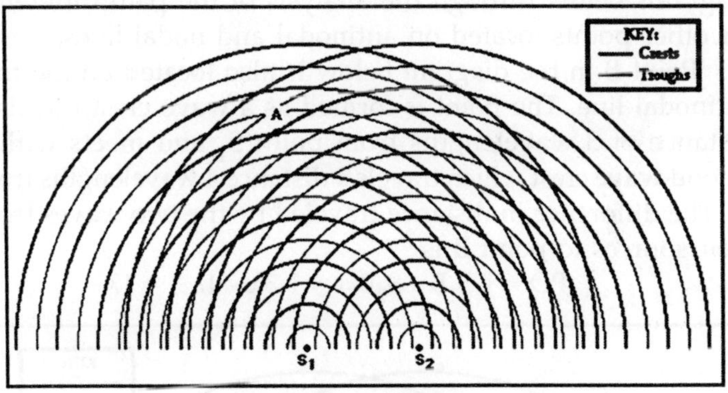

Fig. Point A is a Point on the First Anti-nodal Line to the left of the Central anti-nodal Line

The crest traveling from Source 1 (S_1) travels a distance equivalent to 5 full waves; that is, point A is a distance of 5 wavelengths from Source 1 (S_1).

The crest traveling from Source 2 (S_2) travels a distance equivalent to 6 full waves; point A is a distance of 6 wavelengths from Source 2 (S_2).

While the two wave crests are traveling a different distance from their sources, they meet at point A in such a way that a crest meets a crest. For this specific location on the pattern, the difference in distance traveled (known as the path difference and abbreviated as PD) is

$$PD = | S_1A - S_2A | = | 5\lambda - 6\lambda | = 1\lambda$$

(Note the path difference or PD is the difference in distance traveled by the two waves from their respective sources to a given point on the pattern.)

For point A on the first antinodal line (m =1), the path difference is equivalent to 1 wavelength. But will all points on the first antinodal line have a path difference equivalent to 1 wavelength? And if all points on the first antinodal line have a path difference of 1 wavelength, then will all points on the second antinodal line have a path difference of 2 wavelengths?

And what about the third antinodal line? And what about the nodal lines? These questions are investigated in the diagrams below through the analysis of the path difference for other points located on antinodal and nodal lines.

Point B in the diagram below is also located on the first antinodal line. The point is formed as a wave crest travels a distance of 3 wavelengths from point S_1 and meets with a second wave crest which travels a distance 4 wavelengths from S_2. The difference in distance traveled by the two waves from their sources to point B is

$$PD = | S_1B - S_2B | = | 3\lambda - 4\lambda | = 1\lambda$$

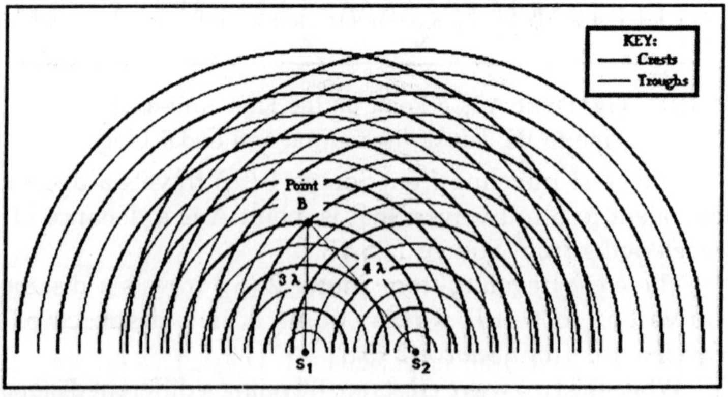

Now we will analyse Point C which is located on the second antinodal line. The point is formed as a wave crest travels a distance of 4 wavelengths from point S_1 and meets with a second wave crest which travels a distance 6 wavelengths from S_2. The difference in distance traveled by the two waves from their sources to point C is

$$PD = | S_1C - S_2C | = | 4\lambda - 6\lambda | = 2\lambda$$

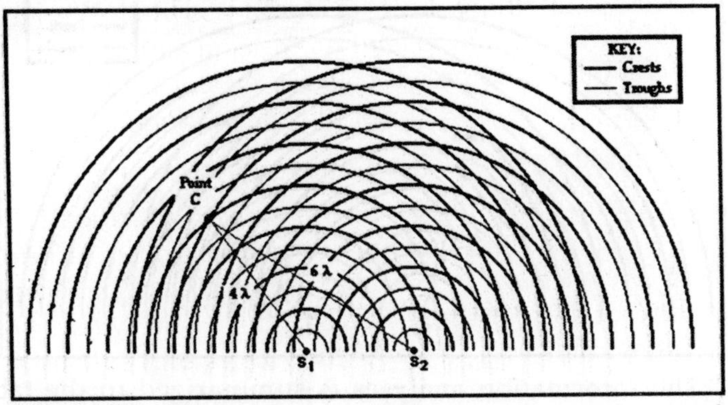

The analysis continues for this same pattern as we explore the path difference for locations on nodal lines. Point D is located on the first nodal line. The point is formed as a wave crest travels a distance of 5 wavelengths from point S_1 and meets with a wave trough which travels a distance 4.5 wavelengths from S_2. The difference in distance traveled by the two waves from their sources to point D is

$$PD = | S_1D - S_2D | = | 5\lambda - 4.5\lambda | = 0.5\lambda$$

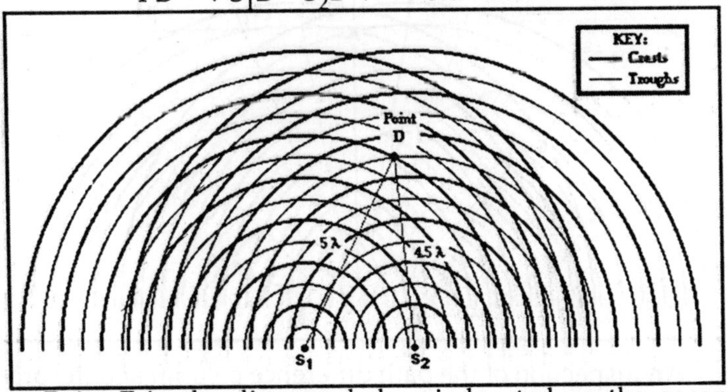

Point E in the diagram below is located on the second nodal line. The point is formed as a wave trough travels a distance of 3.5 wavelengths from point S_1 and meets with a wave crest which travels a distance 5 wavelengths from S_2. The difference in distance traveled by the two waves from their sources to point E is

$$PD = | S_1E - S_2E | = | 3.5\lambda - 5\lambda | = 1.5\lambda$$

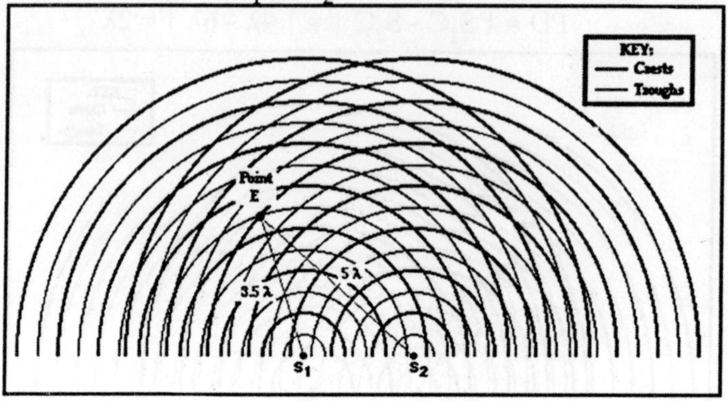

The information analyses is summarized in the table below. Other points on other antinodal and nodal lines are marked on the diagram below and their distance from the sources and their path difference are also summarized in the same table.

Inspect the table and see if we can find a pattern evident in the numbers.

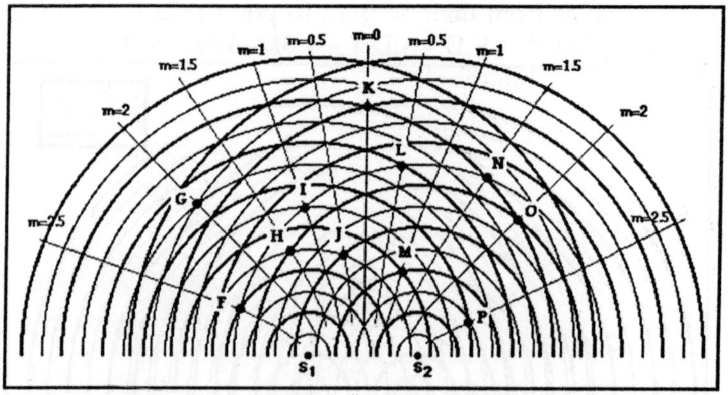

An inspection of the path difference column and the order number column reveals that there is a clear relationship between these two quantities. The path difference is always the order number multiplied by the wavelength. That is,

$$PD = m \bullet \lambda$$

Table. Summary of the Path Difference Analysis

Point	*Antinode*	*Order*	*Distance*	*Distance*	*Path*
		or Node?		*# (m)*	*from S_1(in λ)*
from S_2(in λ)		*Difference(in λ)*			
A	Antinode	1	5 λ	6 λ	1 λ
B	Antinode	1	3 λ	4 λ	1 λ
C	Antinode	2	4 λ	6 λ	2 λ
D	Node	0.5	5 λ	4.5 λ	0.5 λ
E	Node	1.5	3.5 λ	5 λ	1.5 λ
F	Node	2.5	2 λ	4.5 λ	2.5 λ
G	Antinode	2	4.5 λ	6.5 λ	2 λ
H	Node	1.5	2.5 λ	4 λ	1.5 λ
I	Antinode	1	3.5 λ	4.5 λ	1 λ
J	Node	0.5	2.5 λ	3 λ	0.5 λ
K	Antinode	0	6 λ	6 λ	0 λ
L	Node	0.5	5 λ	4 λ	0.5 λ
M	Antinode	1	3 λ	2 λ	1 λ
N	Node	1.5	6 λ	4.5 λ	1.5 λ
O	Antinode	2	6 λ	4 λ	2 λ
P	Node	2.5	4 λ	1.5 λ	2.5 λ

Furthermore, one might notice that the path difference is a whole number of wavelengths for the antinodal positions and a half number of wavelengths for the nodal positions. That is,

Antinodal Points:	$PD = m \bullet \lambda$	where m = 0, 1, 2, 3, 4,...
Nodal Points:	$PD = m \bullet \lambda$	where m = 0.5, 1.5, 2.5, 3.5,...

Explaining the Path Difference Equation

A tedious inspection of a variety of antinodal and nodal points on a typical pattern reveals the above relationships. But why? Why would constructive interference occur when the difference in distance traveled by two waves is equivalent to a whole number of wavelengths?

And why would destructive interference occur when the difference in distance traveled by two waves is equivalent to a half number of wavelengths? To answer these questions, consider the diagrams below. The diagram below shows two waves traveling along different paths from different sources to the same point in such a way that a crest is meeting a crest. Constructive interference will occur at this point. The wave from Source 1 (S_1) travels a distance of 6 wavelengths.

The wave from Source 2 (S_2) travels a distance of 7 wavelengths to reach the same point. The difference in distance traveled by the two waves is one full wavelength. That is, the path difference is 1. When the path difference is one full wavelength, a crest meets a crest and constructive interference occurs.

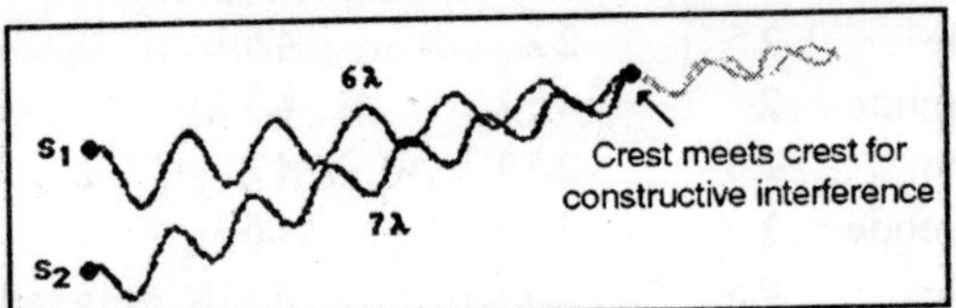

How does a path difference of two wavelengths cause constructive interference? The diagram below shows this. Two waves traveling along different paths from different sources meet at the same point in such a way that a crest is meeting a crest; constructive interference occurs. The wave from Source 1 (S_1) travels a distance of 6 wavelengths.

The wave from Source 2 (S_2) travels a distance of 8 wavelengths to reach the same point. The difference in distance traveled by the two waves is two full wavelengths; that is, the path difference is 2. When the path difference is two full wavelengths, a crest meets a crest and constructive interference occurs.

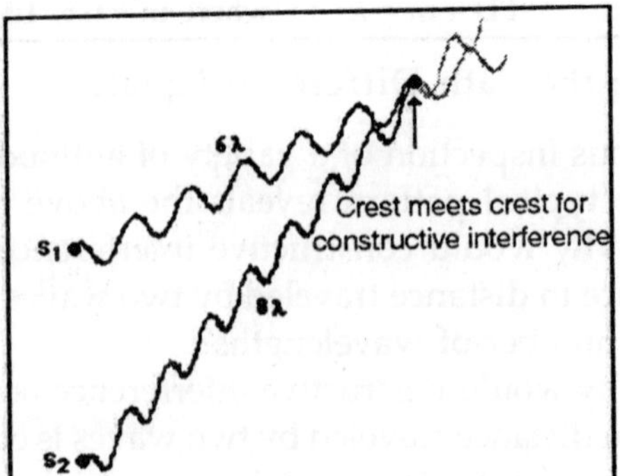

The previous two examples involve the meeting of a crest with a crest. Under what conditions will a trough meet a trough? The diagram below depicts this. Once more, the two waves travel different paths from their different sources and meet in such a manner to constructively interfere; only this time the constructive interference results from a trough meeting a trough.

Observe that the wave from Source 1 (S_1) travels a distance of 6.5 wavelengths. The wave from Source 2 (S_2) travels a distance of 7.5 wavelengths to reach the same point. The difference in distance traveled by the two waves is one full wavelength; that is, the path difference is 1.

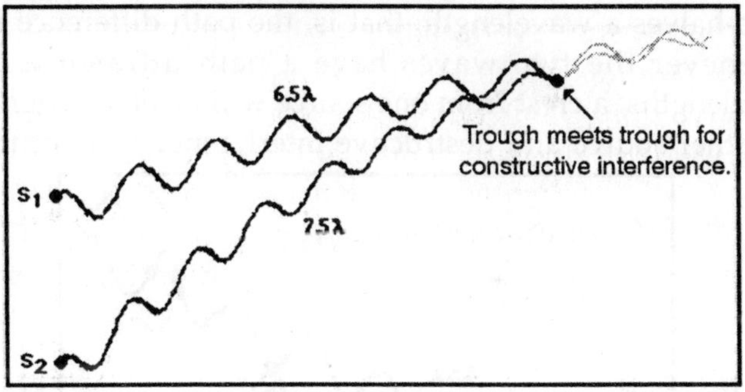

The above examples pertain to the constructive interference which occurs for locations on antinodal lines. In each case, a path difference of a whole number of wavelengths causes a crest to meet a crest or a trough to meet a trough. But what about destructive interference? The diagram below depicts the destructive interference of two waves from the sources.

Observe that the wave from Source 1 (S_1) travels a distance of 3.5 wavelengths. The wave from Source 2 (S_2) travels a distance of 4 wavelengths to reach the same point. The difference in distance traveled by the two waves is one-half a wavelength; that is, the path difference is 0.5. Whenever the two waves have a path difference of one-half a wavelength, a crest from one source will meet a trough from the other source. Destructive interference occurs for path differences of one-half a wavelength.

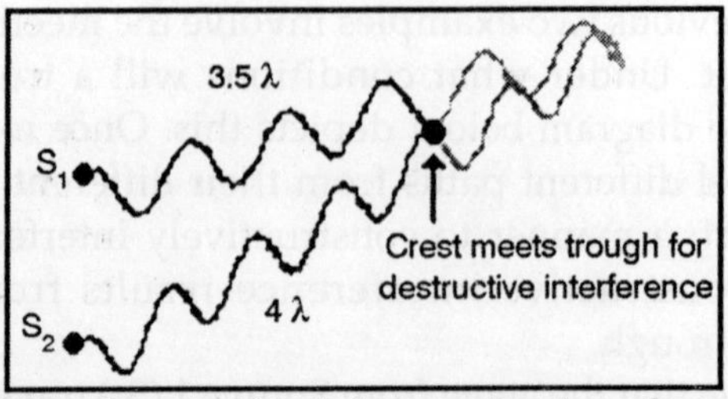

An additional example of destructive interference is shown below. Observe that the wave from Source 1 (S_1) travels a distance of 4.5 wavelengths. The wave from Source 2 (S_2) travels a distance of 6 wavelengths to reach the same point.

The difference in distance traveled by the two waves is three-halves a wavelength; that is, the path difference is 1.5. Whenever the two waves have a path difference of 1.5 wavelengths, a crest from one source will meet a trough from the other source and destructive interference will occur.

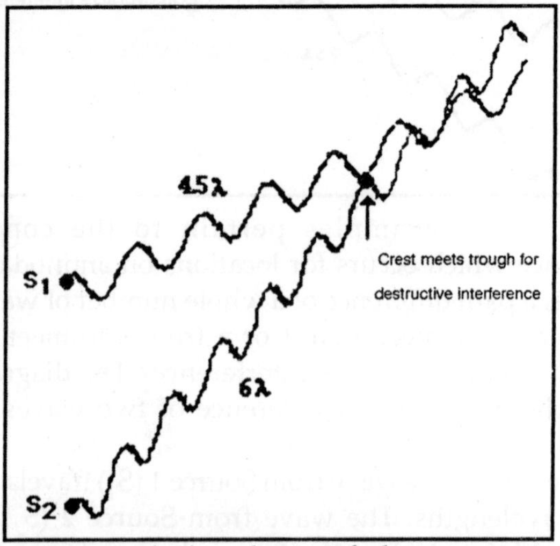

In summary, the analysis of the two-point source interference patterns reveals the following equations:

Antinodal Points:	PD = m •λ	where m = 0, 1, 2, 3, 4,...
Nodal Points:	PD = m •λ	where m =0.5, 1.5, 2.5, 3.5,...

A simple wave model demonstrates why these relationships exist.

Young's Equation

We have seen how circular waves from two sources can interfere in the surrounding space to produce a pattern of antinodes and nodes. The nodal positions are present at locations where destructive interference always occurs and the path difference is a half-number of wavelengths.

The antinodal positions are present at locations where constructive interference always occurs and the path difference is equivalent to a whole number of wavelengths.

Antinodal Points:	PD = m • λ	where m = 0, 1, 2, 3, 4,...
Nodal Points:	PD = m • λ	where m = 0.5, 1.5, 2.5, 3.5,...

The discussion of the interference patterns was introduced by referring to the interference of water waves in a ripple tank. All waves behave the same, whether they are water waves created by vibrating sources in a ripple tank, sound waves produced by two speakers, or light waves produced by two light sources.

For water waves in a ripple tank, the resulting pattern would include locations along the water's surface where water was vibrating up and down with unusually large amplitudes (antinodes). And there would be other locations where the water was relatively undisturbed (nodes).

For sound waves produced by two speakers, the interference pattern would be characterized by locations where the sound intensity was large due to constructive interference (antinodes). And there would be other locations where sound cancellation occurs and the sound intensity was relatively faint or not even heard at all (nodes). But what would be observed in a two-point source light interference pattern?

A Light Interference Pattern

As in any two-point source interference pattern, light waves from two coherent, monochromatic sources (more on coherent and monochromatic later) will interfere constructively and destructively to produce a pattern of antinodes and nodes. Light traveling through the air is typically not seen since there is nothing of substantial size in the air to reflect the light to our eyes. Thus, the pattern formed by light interference cannot be seen unless it is somehow projected onto some form of a screen or a sheet of paper. When light from the two sources is projected onto a screen, the pattern becomes quite evident.

Locations where light constructively interferes corresponds to an abnormally bright spot. Locations where light destructively interferes corresponds to an abnormally dark spot. That is, the antinodes are locations where light from the two individual sources are reinforcing each other and correspond to points of brightness or maximum intensity (sometimes referred to as maxima).

And the nodes are locations where light from the two individual sources are destroying each other and correspond to points of darkness or minimum intensity (sometimes referred to as minima).

Red laser light passing through two narrowly spaced slits is typically used in the classroom to produce this effect. Thus, a pattern of bright red and dark fringes or bands are observed on a screen as shown in the diagram below.

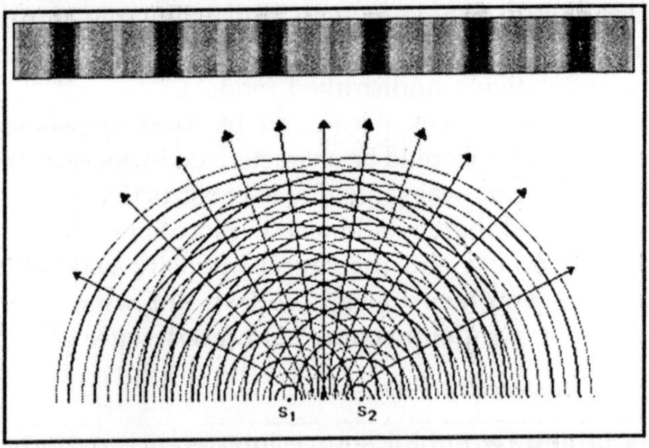

Fig. Screen with Pattern of Bright and Dark Fringes

The central bright band where light displays maximum intensity corresponds to a point on the central antinodal line. The bright bands to the right and the left of the central bright band correspond to the projection of other antinodal lines onto the screen. The dark bands correspond to the projection of the nodal lines onto the screen.

Each antinodal and nodal line is assigned a number or order value (m). The red band of maximum brightness located in the centre of screen (the central maximum) is assigned an order number of m = 0. The other bright red bands to the left and the right of the central maximum are assigned whole number values of 1, 2, 3,... as shown in the diagram below. The dark bands on the pattern are assigned half number values of 0.5, 1.5, 2.5,... as shown in the diagram below.

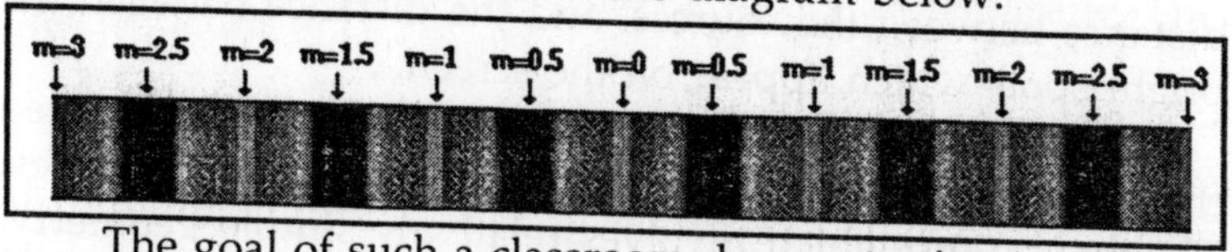

The goal of such a classroom demonstration is typically twofold:

- To demonstrate the wavelike nature of light by displaying its ability to interfere
- To use the interference pattern to measure the wavelength of light and verify the mathematical model of two-point source interference.

Derivation of Young's Equation

In 1801, this experiment was performed for the first time by Thomas Young. Young expanded the mathematical model presented above by relating the wavelength of light to observable and measurable distances. Today, an experimental setup similar to that of Thomas Young's is commonly used in a Physics classroom to repeat the experiment and to measure the wavelength of light.

In the experiment as it is commonly performed today, light from a laser beam is passed through two narrowly-spaced slits in a slide or sheet of paper. The light diffracts through the slits and interfere in the space beyond the slits. Thus, the slits serve as the two sources. The interference pattern is then projected onto a screen, paper or a whiteboard located several meters away. The spatial separation of nodes and antinodes on the screen is clearly seen.

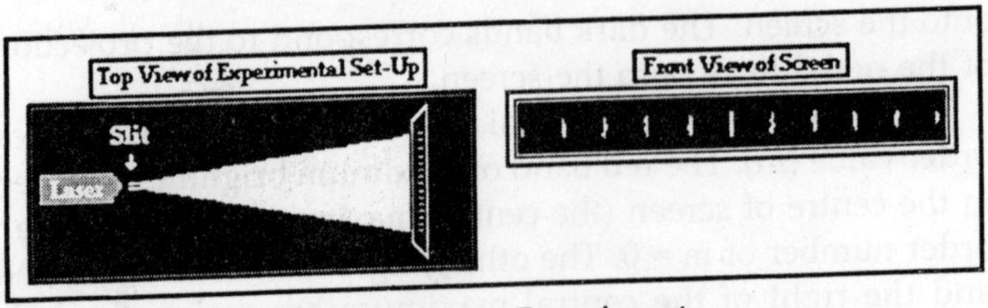

Fig. Young's Experiment

The most reliably measured distances in this experimental procedure are the distance from the sources to the screen, the distance between the sources, and the distance between the bright spots which appear on the screen.

Thus, Thomas Young derived an equation which related the wavelength of the light to these measurable distances. The derivation, which involves relatively simple geometry, right-angle trigonometry and algebra, is repeated below.

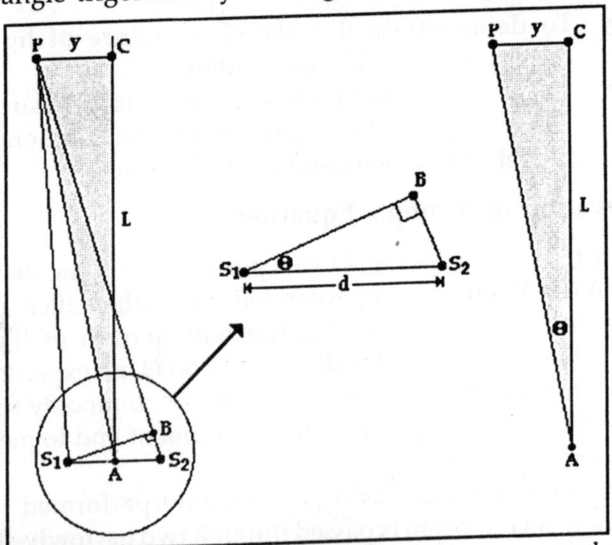

The diagram below on the left depicts two sources labeled S_1 and S_2 and separated by some distance d.

Point P is a point on the screen which happens to be located on some nodal or antinodal line; as such, there is an order value (m) associated with this point. Point C is the central point on the screen.

The distance from point P to point C as measured perpendicular to the central antinodal line will be referred to as y. The screen is located a distance of L from the sources. In the following derivation, the wavelength of light will be related to the quantities d, m, y and L.

On the diagram above, source S_2 is further from point P than source S_1 is. The extra distance traveled by waves from S_2 can be determined if a line is drawn from S_1 perpendicular to the line segment S_2P. This line is drawn in the diagram on the left above; it intersects line segment S_2P at point B.

If point P (a bright spot on the screen) is located a great distance from the sources, then it follows that the line segment S_1P is the same distance as BP. As such, the line segment S_2B is simply the path difference. That is, the small distance S_2B is equal to the difference in distance traveled by the two waves from their individual sources to point P on the screen. The logic is as follows:

	Assertion	Logic/Rationale
i	$S_2P = S_2B + BP$	See diagram above left
ii	$PS_1B = PBS_1$	If screen is very far away (L >>> y), then lines S_1P and BP are ||
iii	$S_1P = BP$	If S_1P and BP are || and line S_1B is perpendicular to BP, then the length BP = length S_1P
iv	$S_2P = S_2B + S_1P$	Since S_1P = BP, the expression S_1P
can		be substituted into i
v	$S_2B = S_2P - S_1P$	Algebraic manipulation of iv;
subtract		S_1P from each side
vi	S_2B is the PD	The path difference (PD) is defined
as		$\| S_2P - S_1P \|$

Note that step ii in the logical proof above demanded that an assumption be made: the screen must be very far away compared to the spacing between point P and the central antinodal line. That is, L >>> y. This is an assumption which underlies Young's derivation of his wavelength equation. The equation is only as valid as this assumption is true.

The yellow triangle in the diagram on the left above is

enlarged and redrawn in the middle of the graphic. The triangle is a right triangle with an angle theta and a hypotenuse of d. Using the sine function, it can be stated that

$$\text{Sine}(\theta) = S_2B / d$$

But since it has been previously stated that the path difference (PD) is equal to the length of the line segment S_2B, the above equation can be rewritten as

$$\text{Sine}(\theta) = PD / d$$

It can be further asserted that the pink triangle (ΔS_1BS_2) and the yellow triangle (ΔACP) in the diagrams above are similar triangles. To prove that any two triangles are similar, one must show that they have two corresponding angles which are equal. Since the line segment PC was drawn perpendicular to the central antinodal line, it forms a 90 degree angle with the line AC. Thus, the corresponding angles S_1BS_2 and ACP are equal.

The second set of corresponding angles of equal measure are $\angle S_1S_2B$ and APC. This can be proven by returning to the assumption that the screen is very far away (L >>> y). Clearly, the lines S_1S_2 and PC are parallel lines. As such, the diagonal line S_2P creates two alternating angles which are equal in measure - that is, $\angle S_2PC = \angle S_1S_2B$.

Since it is assumed that L >>> y, the line segment S_2P and AP are roughly parallel to each other and thus make roughly the same angle with the line segment S_1S_2. With this assumption, it is thus proven that $\angle APC = \angle S_2PC$. Since the $\angle S_2PC$ is equal to both $\angle S_1S_2B$ and $\angle APC$, it follows that $\angle S_1S_2B = \angle APC$. As such, the pink triangle (ΔS_1BS_2) and the yellow triangle (ΔACP) have two corresponding angles which are equal and thus are similar triangles.

If the pink triangle (ΔS_1BS_2) and the yellow triangle (ΔACP) are similar, all corresponding angles are equal in measure, and so $\angle S_2S_1B = \angle PAC$. These two angles are labeled as theta in the diagrams above. The use of trigonometric functions allows one to relate the angle theta to the reliably measured distances d, y and L.

Sine θ = PD / d	Tangent θ = y / L

The above logic has consistently assumed that the screen

upon which the interference pattern is projected is very far away; that is, L >>> y. This is typically the case for visible light interference patterns. In fact, the L value is typically on the order of several meters while the y value is on the order of a couple of centimeters.

For such dimensions, the angle theta is less than 1 degree. For such small angles, the approximation that the sine = tangent can be made. Taking 1 degree as a sample angle, calculated values of the sine and tangent can be compared.

Sine (1 deg) = 0.017452406 Tangent (1 deg) = .017455064

Note that the values for the sine and the tangent of 1 degree show agreement out to the fourth significant digit. Since the sine and the tangent of these small angles are approximately equal, we can state that their ratio of lengths is also equal. That is,

$$PD / d = y / L$$

It was shown that the path difference (PD) for any point on the pattern is equal to m •, where m is the order number of that point and is the wavelength. By substitution,

$$m \bullet \lambda / d = y / L$$

As a final step in the derivation, the equation can be algebraically manipulated so that the wavelength () is by itself:

$$\lambda = y \bullet d / (m \bullet L)$$

The wavelength of laser light can be experimentally determined by selecting a point (referred to as point P) on a nodal and antinodal line of known order value (m) and making the following measurements:

- the distance between the slits or sources of the two light waves (d)
- the perpendicular distance from the point P to a point on the central antinodal line (y)
- the distance from point P to the sources (L)

The Importance of Coherent Light Sources

Visible light waves - those which humans can see - have an abnormally short wavelength. For instance, red light has a wavelength of about 650 nanometers. Since there are one billion nanometers in a meter, and one thousand millimeters in a meter, the wavelength of red light is less than one-thousandth of a millimeter. That's a very short wave. And being a short wave, the distance between positions of constructive interference and positions of destructive interference is very small. Thus, the effects of interference for visible light waves is difficult to observe.

Complicating the task of observing the interference of visible light waves is the fact that light from the two sources must be coherent. Two light sources which maintain a constant phase difference with each other are said to be coherent light sources. Light visible to the human eye makes a complete cycle of *vibration* from crest to trough and back to crest in roughly 10^{-15} seconds. If we think of a light wave as a transverse wave pattern with crests and troughs, then a crest is typically created every 10^{-15} seconds.

Consider two light sources producing light waves at the same frequency, but one source is creating a crest just prior to the moment in time when the other source is creating a crest. Such light sources are not at the same phase in their cycle of light production. They are said to be out of phase.

Yet if they maintain the same difference in phase, they are considered coherent light sources. Even if the sources of light do not stay *in step* with each other, as long as the amount by which they are out of step remains the same over time, the light sources are said to be coherent.

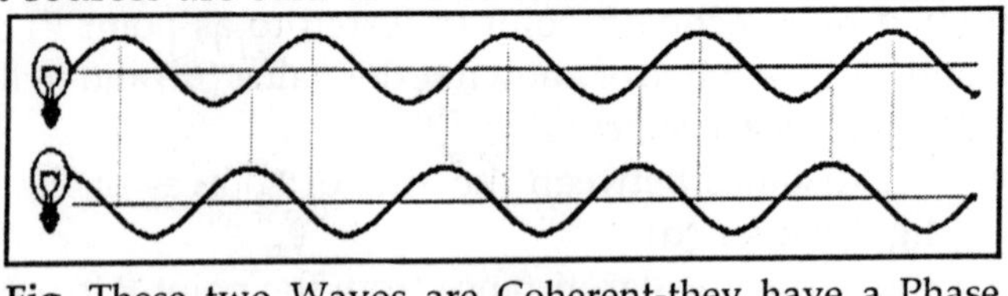

Fig. These two Waves are Coherent-they have a Phase difference which is Constant Over Time

Typical light sources such as incandescent light bulbs have an intrinsic irregularity associated with the manner in which they produce light. The production of light results when atoms of the filaments or source become electrically excited and produce an electromagnetic or light wave. Since these excitations occur in an unpredictably random fashion, the light waves which they produce do not maintain a constant phase difference with each other.

Approximately once every 10^{-8} seconds, a source will randomly alter its phase. While waves from these sources will interfere to produce an interference pattern, the lack of coherence means that the pattern will constantly change with time. In fact, the patterns change so rapidly with time (once every 10^{-8} seconds) that there is no possibility that the human eye could ever discern a pattern. To produce a stable and unchanging interference pattern, it is essential that the two sources be coherent.

Young's Experiment

Thomas Young's effort to derive an equation relating the wavelength of a light source to reliably measured distances associated with a two-point source light interference pattern. The equation, known as Young's equation is:

$$\lambda = y \bullet d / (m \bullet L)$$

Young devised and performed an experiment to measure the wavelength of light. It was important that the two sources of light which form the pattern be coherent. The difficulty confronting Young was that the usual light sources of the day (candles, lanterns, etc.) could not serve as coherent light sources. Young's method involved using sunlight that entered the room through a pinhole in a window shutter.

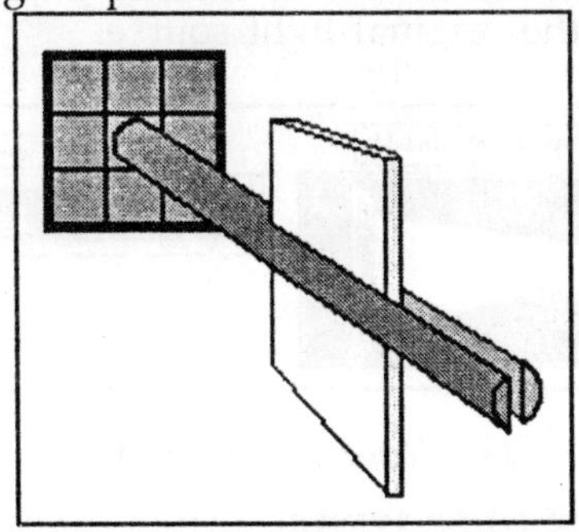

Fig. A Single Pinhold Beam of Sunlight is Split Into two Coherent Beams

A mirror was used to direct the pinhole beam horizontally

across the room. To obtain two sources of light, Young used a small paper card to break the single pinhole beam into two beams, with part of the beam passing by the left side of the card and part of the beam passing by the right side of the card.

Since these two beams emerged from the same source - the sun - they could be considered coming from two coherent sources. Light waves from these two sources (the left side and the right side of the card) would interfere. The interference pattern was then projected onto a screen where measurements could be made to determine the wavelength of light.

Today's classroom version of the same experiment is typically performed using a laser beam as the source. Rather than using a note card to split the single beam into two coherent beams, a carbon-coated glass slide with two closely-spaced etched slits is used.

The slide with its slits is most commonly purchased from a manufacturer who provides a measured value for the slit separation distance - the d value in Young's equation. Light from the laser beam diffracts through the slits and emerges as two separate coherent waves.

The interference pattern is then projected onto a screen where reliable measurements can be made of L and y for a given bright spot with order value m. Knowing these four values allows a student to determine the value of the wavelength of the original light source.

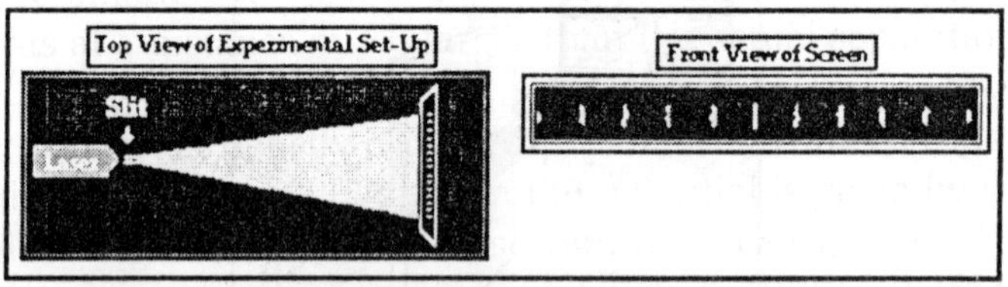

Fig. Young's Experiment

To illustrate some typical results from this experiment and the subsequent analysis, consider the sample data provided below for d, y, L and m.

The determination of the wavelength demands that the

above values for d, y, L and m be substituted into Young's equation.

$$\lambda = y \bullet d / (m \bullet L)$$

Table. Data Table

Slit Separation (d)	0.250 mm
Distance from Slits to Screen (L)	9.78 m
Distance from AN_0 to AN_4 (y)	10.2 cm
Order value (m)	4

(Note: AN_0 = central antinode and AN_4 = fourth antinode)

Careful inspection of the units of measurement is always advisable. The sample data here reveal that each measured quantity is recorded with a different unit. Before substituting these measured values into the above equation, it is important to give some thought to the treatment of units. One means of resolving the issue of nonuniform units is to simply pick a unit of length and to convert all quantities to that unit.

If doing so, one might want to pick a unit that one of the data values already has so that there is one less conversion. A wise choice is to chose the meter as the unit to which all other measured values are converted. Since there are 1000 millimeters in 1 meter, the 0.250 mm is equivalent to 0.000250 meter. And since there are 100 centimeters in 1 meter, the 10.2 cm is equivalent to 0.102 m. Thus, the new values of d, y and L are:

d	0.000250 m
L	9.78 m
y	0.102 m
m	4

While the conversion of all the data to the same unit is not the only means of treating such measured values, it might be the most advisable - particularly for those students who are less at ease with such conversions.

Now that the issue regarding the units of measurement have been resolved, substitution of the measured values into Young's equation can be performed.

$$\lambda = (0.102 \text{ m}) \bullet (0.000250 \text{ m}) / [(4) \bullet (9.78 \text{ m})]$$

$$\lambda = 6.52 \times 10^{-7} \text{ m}$$

As is evident here, the wavelength of visible light is rather small. For this reason wavelength is often expressed using the unit nanometer, where 1 meter is equivalent to 10^9 nanometers. Multiplying by 10^9 will convert the wavelength from meters to nanometers (abbreviated nm).

$\lambda = 652$ nm

Other Applications of Two-Point Source Interference

As is often the case, physics is not a science which is restricted to the sterile confines of a laboratory. Physics is naturally and frequently seen by any observer who glances at the world around them. An informed physics student should *see* physics in action on a daily basis and should be able to exclaim without embarrassment - "that happens because of physics."

These physics concepts and principles are simply humankind's attempt to explain the observable. So where in this world do we observe two-point source interference? Where can we experience the phenomenon that light taking two paths from two locations to the same point in space can undergo constructive and destructive interference?

THE BIG IDEA

Before identifying and explaining real-world examples of two-point source interference, the big idea behind it ought to be reviewed. It was noted that two coherent waves traveling along two different paths to the same point will interfere constructively if there is a difference in distance traveled that is equivalent to a whole number of wavelengths.

And similarly, two coherent waves traveling along two different paths to the same point will interfere destructively if there is a difference in distance traveled that is equivalent to a half number of wavelengths. The difference in distance traveled by the waves from the two sources to the single point is referred to as the path difference. That is,

Constructive Interference: $PD = m \bullet \lambda$ where $m = 0, 1, 2, 3, 4,...$

Destructive Interference: $PD = m \bullet \lambda$ where $m = 0.5, 1.5, 2.5, 3.5,...$

These principles were presented to explain the two-point source interference patterns that are characteristic of Young's experiment and a wavelength measurement. Yet, these principles are more general in the sense that they can explain any physical situation in which waves take two different paths from two coherent sources to the same point.

Such coherent waves will undergo interference. And if the difference in distance traveled is a whole number of wavelengths, then the interference will occur in such a way that the two waves will constructively reinforce each other. Likewise, if the difference in distance traveled is a half number of wavelengths, then the interference will occur in such a way that the two waves will destroy each other.

THE APPLICATIONS

The clue to finding two-point source interference in the real world would be to look for situations in which waves from two coherent sources travel along two different paths to the same point. Since the two waves must be coherent, it is best that they can be traced to the same source, but separated into two paths at some point due to passage through two openings or reflection off a barrier.

A common example of this involves the interference of radio wave signals which occur at the antenna of a home when radio waves from a very distant transmitting station take two different paths from the station to the home. This is relatively common for homes located near mountain cliffs. In such an instance, waves travel directly from the transmitting station to the antenna and interfere with other waves which reflect off the mountain cliffs behind the home and travel back to the antenna.

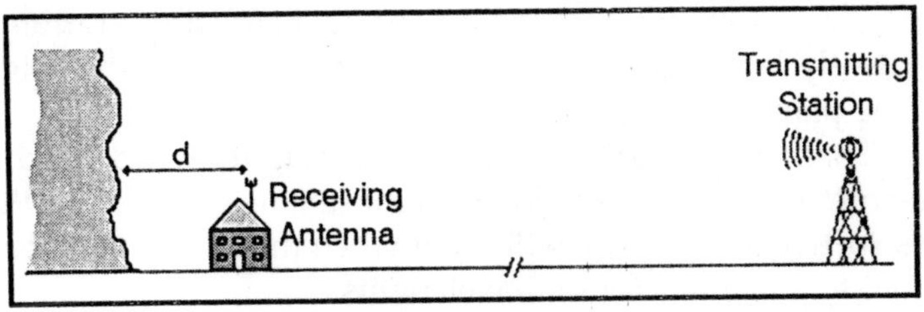

In this case, waves are taking two different paths from the source to the antenna - a direct path and a reflected path. Clearly, each path is represented by a different distance traveled from the source to the home, with the reflected pathway corresponding to the longer distance of the two.

If the home is located some distance d from the mountain cliffs, then the waves which take the reflected path to the home will be traveling an extra distance given by the expression 2•d. The 2 in this expression is due to the fact that the waves taking the reflected path must travel past the antenna to the cliffs (a distance d) and then back to the antenna from the cliff (a second distance d).

Thus, the path difference of 2•d results in destructive interference whenever it is equal to a *half number* of wavelengths.

Destructive:	$PD = 2 \cdot d = m \cdot \lambda$	where m = 0.5, 1.5, 2.5,...

Since radio stations transmit their signals at a specific and known frequency, the wavelengths of these light waves can be determined by relating it to the transmitted frequency and the light speed (3×10^8 m/s).

This same principle of destructive interference of radio signals can be observed when waves from the transmitting source reflect off airplanes which are flying overhead. In this case, there will be a difference in the distance traveled by the wave moving along the direct path to the antenna and the wave which travels along the reflected path off the plane to the antenna.

While the interference is momentary (the plane does not remain in a stationary location), it is nonetheless observable. If we suppose that the plane is directly overhead and that the distance from the antenna to the transmitting station is relatively large, then the path difference is simply the height of the plane above the house.

There are a variety of potential heights which lead to destructive interference. Each height satisfies the criteria that destructive interference will occur when the path difference is equal to a *half number* of wavelengths.

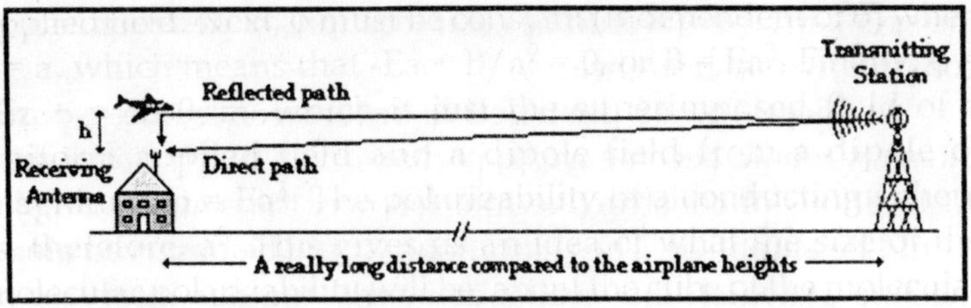

Destructive: PD = height = m • λ where m = 0.5, 1.5, 2.5,...

By substituting various values of m into the equation, the variety of potential heights can be determined.

A final application of two-point source interference which is discussed here involves the interference of sound waves. All waves, whether light waves, sound waves or water waves, exhibit the same characteristics properties. Waves refract, reflect, diffract and interfere in the same manner according to the same rules. And as such, two coherent sound waves traveling along different paths to the same point will destructively interfere provided that the path difference is equal to a *half number* of wavelengths.

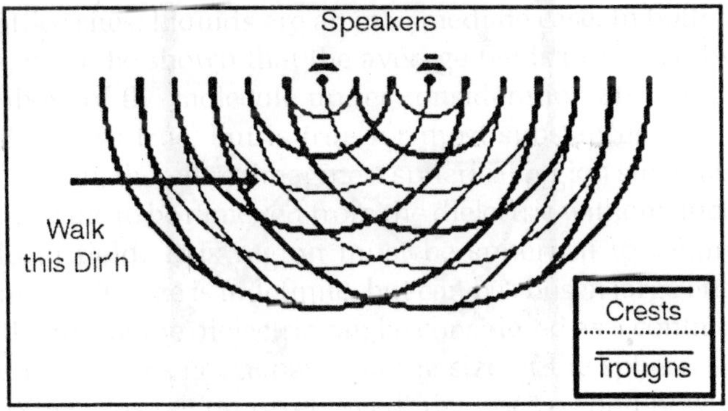

Fig. Sound Interference Observed in a Large Room

A relatively common demonstration of sound wave interference can be performed with two speakers in a large room such as an auditorium. If both speakers are hooked up to the same sound source producing a monotone sound, then a sound interference pattern can be observed within the room. If one were to walk along a line parallel to the line connecting the speakers, there would be clear locations of destructive and constructive interference.

At locations of destructive interference, the sound intensity would become weak, perhaps even barely noticeable. At locations of constructive interference, the sound intensity would be amplified. These locations would be observed along the line at which one walks at nearly regular intervals of distance.

Imagine that there were simply two speakers in a large auditorium set up so that there were certain seats which were located along nodal lines for particular frequencies.

When those particular frequencies were sounded out by the speakers, the people in the seats along those nodal lines would be at locations of destructive interference and would not hear the sound from the speakers (at least during the time that those particular frequencies were being sounded out).

Acoustic engineers must take these factors into account when designing auditorium walls and ceilings.

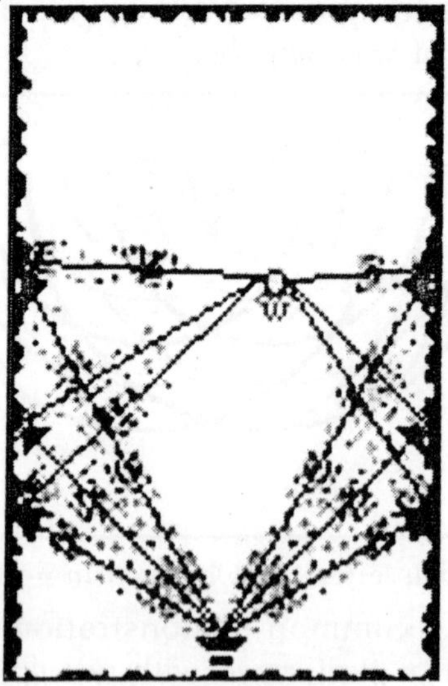

Fig. When Designed Correctly, Auditorium Walls will Reflect Sound in such a way to Give the Room a Full and Lively Feel

The walls and ceilings must act as additional "sources" of sound as they reflect sound waves to all parts of the room so that even when waves traveling directly from the speakers to a seat undergo destructive interference, there is still sound reflecting off walls and ceilings to the same seats. In this sense, the walls and ceilings of a a well designed auditorium serve to reflect sound in such a way as to *fill in* those locations where destructive interference might be occurring.

Light From Two Light Bulbs Not Form an Interference Pattern

When the topic of Young's experiment and interference is discussed, the question is often raised: Why doesn't light from two light bulbs undergo interference to produce a two-source interference pattern? Why do I not observe bright and dark fringes along my living room wall from the interference of light from two lamps? The explanation pertains to the nature of ordinary incandescent bulbs and to the necessity of coherent sources. The importance of coherent light sources was emphasized. Coherent light sources are sources of light which produce waves which have a constant phase difference between them over a significant duration of time.

Two waves may have the same frequency and wavelength but be offset from each other in that they are at different points in a complete cycle. For instance, one wave could be at a crest position just prior to the moment in time when the other source is creating a crest. They are said to be out of phase. Yet if they maintain the same difference in phase, they are considered coherent light sources.

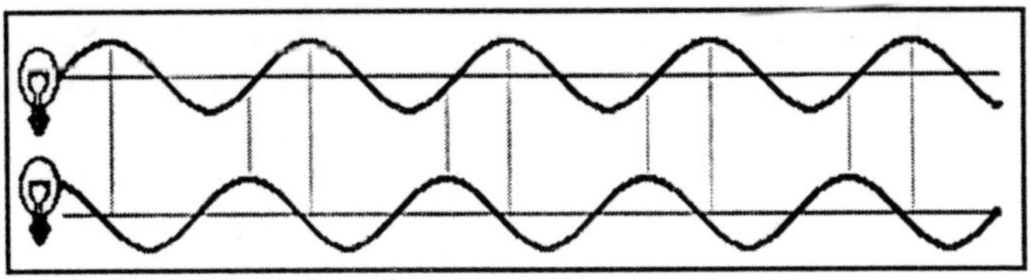

Fig. These Two Waves are Coherent–they have a Phase Difference which is Constant over Time

Even if the sources of light do not stay *in step* with each other, as long as the amount by which they are out of step remains the same over time, the light sources are said to be coherent. To be coherent, two waves must have the same frequency and there must not be any disruption of their cycle over the course of time.

In an incandescent bulb, the vibrations of electrons within the filament lead to the production of the electromagnetic wave. Several million times in a second, there are small disruptions of the filament which result in discontinuities in the waves which they produce.

Ultimately, two different incandescent bulbs are unable to produce light waves which maintain a constant difference in phase over time. As a result, any interference pattern that does occur will not endure for a sufficient length of time to allow the human eye to observe the pattern.

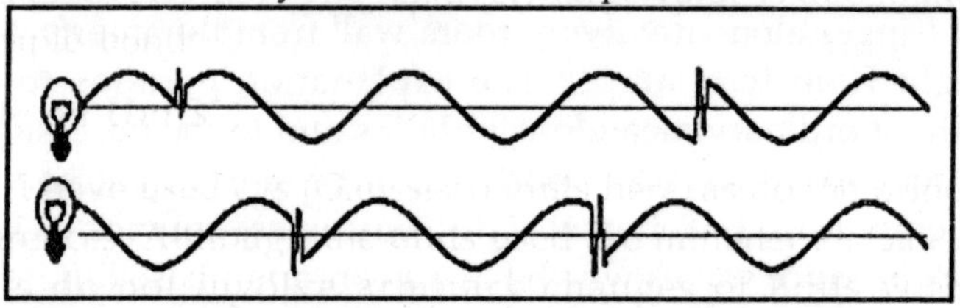

Fig. These Two Waves are NOT Coherent–the Phase Difference Fluctuates Over Time

Chapter 6

Polarization

Polarization is a property of waves that describes the orientation of their oscillations. For transverse waves such as many electromagnetic waves, it describes the orientation of the oscillations in the plane perpendicular to the wave's direction of travel. The oscillations may be oriented in a single direction (linear polarization), or the oscillation direction may rotate as the wave travels (circular or elliptical polarization).

For longitudinal waves such as sound waves in fluids the direction of oscillation is by definition along the direction of travel (i.e. polarization is not possible). Guided modes in waveguides and optical fibers can carry waves with both transverse and longitudinal oscillations. Such waves do have polarization.

Polarization is used in areas of science and technology dealing with wave propagation, such as optics, seismology, and telecommunications. For electromagnetic waves such as light, the polarization is described by specifying the direction of the wave's electric field. According to the Maxwell equations, the direction of the magnetic field is uniquely determined for a specific electric field distribution and polarization. For transverse sound waves in a solid, the polarization is associated with the direction of the shear stress in the plane perpendicular to the propagation direction.

PLANE WAVES

The simplest manifestation of polarization to visualize is that of a plane wave, which is a good approximation of most light waves (a plane wave is a wave with infinitely long and wide wavefronts). For plane waves the transverse condition requires that the electric and magnetic field be perpendicular to the direction of propagation and to each other.

Conventionally, when considering polarization, the electric field vector is described and the magnetic field is ignored since it is perpendicular to the electric field and proportional to it. The electric field vector of a plane wave may be arbitrarily divided into two perpendicular components labeled x and y (with z indicating the direction of travel).

For a simple harmonic wave, where the amplitude of the electric vector varies in a sinusoidal manner in time, the two components have exactly the same frequency. However, these components have two other defining characteristics that can differ. First, the two components may not have the same amplitude. Second, the two compcnents may not have the same phase, that is they may not reach their maxima and minima at the same time. Mathematically, the electric field of a plane wave can be written as,

$$\vec{E}(\vec{r},t) = \mathrm{Re}\left[\left(A_x, A_y \cdot e^{i\phi}, 0\right)e^{i(kz-\omega t)}\right]$$

or alternatively,

$$\vec{E}(\vec{r},t) = \left(A_x \cdot \cos(kz-\omega t), A_y \cdot \cos(kz-\omega t+\phi), 0\right)$$

where A_x and A_y are the amplitudes of the x and y directions and φ is the relative phase between the two components. The shape traced out in a fixed plane by the electric vector as such a plane wave passes over it (a Lissajous figure) is a description of the polarization state.

The following figures show some examples of the evolution of the electric field vector (blue), with time(the vertical axes), at a particular point in space, along with its x and y components (red/left and green/right), and the path traced by the tip of the vector in the plane (purple): The same evolution would occur when looking at the electric field at a particular time while evolving the point in space, along the direction opposite to propagation.

The two orthogonal (perpendicular) components are in phase. In this case the ratio of the strengths of the two components is constant, so the direction of the electric vector (the vector sum of these two components) is constant. Since the tip of the vector traces out a single line in the plane, this special case is called linear polarization. The direction of this line depends on the relative amplitudes of the two components.

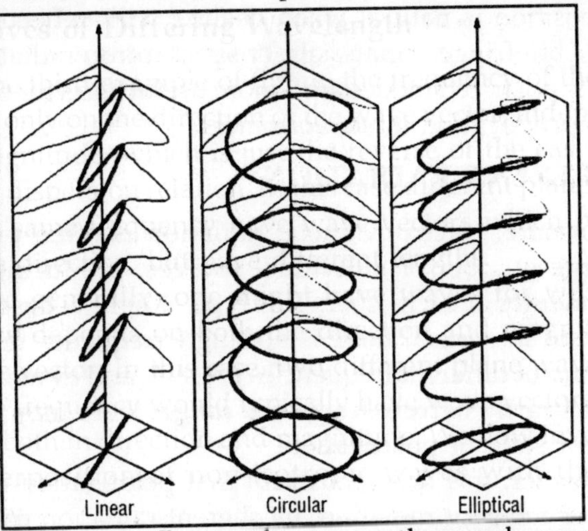

In the middle figure, the two orthogonal components have exactly the same amplitude and are exactly ninety degrees out of phase. In this case one component is zero when the other component is at maximum or minimum amplitude. There are two possible phase relationships that satisfy this requirement: the x component can be ninety degrees ahead of the y component or it can be ninety degrees behind the y component.

In this special case the electric vector traces out a circle in the plane, so this special case is called circular polarization. The direction the field rotates in, depends on which of the two phase relationships exists. These cases are called *right-hand circular polarization* and *left-hand circular polarization,* depending on which way the electric vector rotates.

In all other cases, where the two components are not in phase and either do not have the same amplitude and/or are not ninety degrees out of phase, the polarization is called elliptical polarization because the electric vector traces out an ellipse in the plane (the *polarization ellipse*). The "Cartesian" decomposition of the electric field into x and y components is, of course, arbitrary. Plane waves of any polarization can be described instead by combining waves of opposite circular polarization, for example. The Cartesian polarization decomposition is natural when dealing with reflection from surfaces, birefringent materials, or synchrotron radiation. The circularly polarized modes are a more useful basis for the study of light propagation in stereoisomers.

INCOHERENT RADIATION

In nature, electromagnetic radiation is often produced by a large number of individual sources, producing waves independently of each other. This type of light is described as incoherent. In general there is no single frequency but rather a spectrum of different frequencies present, and even if filtered to an arbitrarily narrow frequency range, there may not be a consistent state of polarization.

However, this does not mean that polarization is only a feature of coherent radiation. Incoherent radiation may show statistical correlation between the components of the electric field, which can be interpreted as partial polarization. In general it is possible to describe an observed wave field as the sum of a completely incoherent part (no correlations) and a completely polarized part. One may then describe the light in terms of the degree of polarization, and the parameters of the polarization ellipse.

Parameterizing Polarization

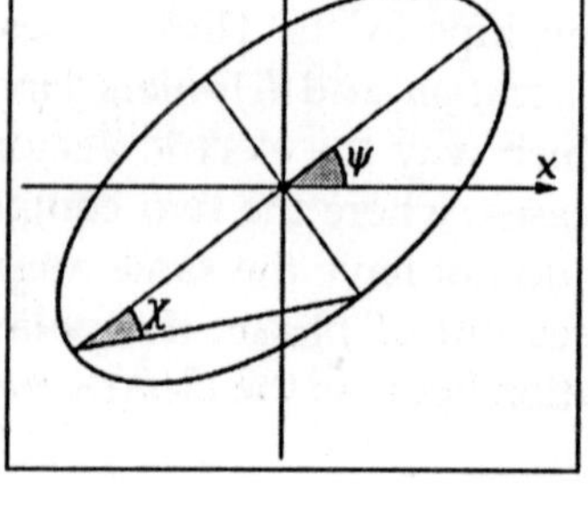

For ease of visualization, polarization states are often specified in terms of the polarization ellipse, specifically its orientation and elongation. A common parameterization uses the azimuth angle, Ψ (the angle between the major semi-axis of the ellipse and the *x*-axis) and the ellipticity, ε (the ratio of the two semi-axes).

An ellipticity of zero corresponds to linear polarization and an ellipticity of 1 corresponds to circular polarization. The arctangent of the ellipticity, $\chi = \arctan \varepsilon$ (the "ellipticity angle"), is also commonly used. An example is shown in the diagram to the right. An alternative to the ellipticity or ellipticity angle is the eccentricity, however unlike the azimuth angle and ellipticity angle, the latter has no obvious geometrical interpretation in terms of the Poincaré sphere.

Full information on a completely polarized state is also provided by the amplitude and phase of oscillations in two components of the electric field vector in the plane of polarization.

This representation used to show how different states of polarization are possible. The amplitude and phase information can be conveniently represented as a two-dimensional complex vector:

$$e = \begin{bmatrix} \alpha_1 e^{i\theta_1} \\ \alpha_2 e^{i\theta_2} \end{bmatrix}.$$

Here a_1 and a_2 denote the amplitude of the wave in the two components of the electric field vector, while θ_1 and θ_2 represent the phases. The product of a Jones vector with a complex number of unit modulus gives a different Jones vector representing the same ellipse, and thus the same state of polarization. The physical electric field, as the real part of the Jones vector, would be altered but the polarization state itself is independent of absolute phase.

The basis vectors used to represent the Jones vector need not represent linear polarization states (i.e. be real). In general any two orthogonal states can be used, where an orthogonal vector pair is formally defined as one having a zero inner product. A common choice is left and right circular polarizations, for example to model the different propagation of waves in two such components in circularly birefringent media or signal paths of coherent detectors sensitive to circular polarization.

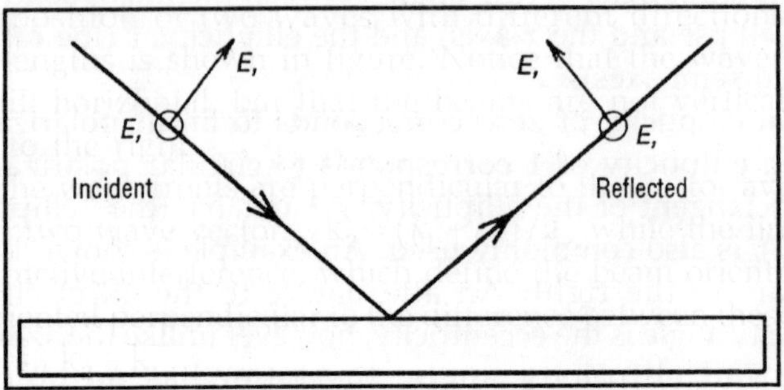

Reflection of a plane wave from a surface perpendicular to the page. The p-components of the waves are in the plane of the page, while the s components are perpendicular to it.

Regardless of whether polarization ellipses are represented using geometric parameters or Jones vectors, implicit in the parameterization is the orientation of the coordinate frame. This permits a degree of freedom, namely rotation about the propagation direction. When considering light that is propagating parallel to the surface of the Earth, the terms "horizontal" and "vertical" polarization are often used, with the former being associated with the first component of the Jones vector, or zero azimuth angle.

On the other hand, in astronomy the equatorial coordinate system is generally used instead, with the zero azimuth (or position angle, as it is more commonly called in astronomy to avoid confusion with the horizontal coordinate system) corresponding to due north. Another coordinate system frequently used relates to the plane made by the propagation direction and a vector normal to the plane of a reflecting surface.

This is known as the *plane of incidence*. The component of the electric field parallel to this plane is termed *p-like* (parallel) and the component perpendicular to this plane is termed *s-like.* Light with a p-like electric field is said to be *p-polarized, pi-polarized, tangential plane polarized,* or is said to be a *transverse-*

magnetic (TM) wave. Light with an s-like electric field is *s-polarized,* also known as *sigma-polarized* or *sagittal plane polarized,* or it can be called a *transverse-electric* (TE) wave.

In the case of partially-polarized radiation, the Jones vector varies in time and space in a way that differs from the constant rate of phase rotation of monochromatic, purely-polarized waves. In this case, the wave field is likely stochastic, and only statistical information can be gathered about the variations and correlations between components of the electric field. This information is embodied in the coherency matrix:

$$\begin{aligned}\Psi &= \left\langle \mathrm{ee}^{\dagger} \right\rangle \\ &= \left\langle \begin{bmatrix} e_1 e_1^{*} & e_1 e_2^{*} \\ e_2 e_1^{*} & e_2 e_2^{*} \end{bmatrix} \right\rangle \\ &= \left\langle \begin{bmatrix} \alpha_1^2 & \alpha_1 \alpha_2 e^{i(\theta_1 - \theta_2)} \\ \alpha_1 \alpha_2 e^{-i(\theta_1 - \theta_2)} & \alpha_2^2 \end{bmatrix} \right\rangle\end{aligned}$$

where angular brackets denote averaging over many wave cycles. Several variants of the coherency matrix have been proposed: the Wiener coherency matrix and the spectral coherency matrix of Richard Barakat measure the coherence of a spectral decomposition of the signal, while the Wolf coherency matrix averages over all time/frequencies.

The coherency matrix contains all of the information on polarization that is obtainable using second order statistics. It can be decomposed into the sum of two idempotent matrices, corresponding to the eigenvectors of the coherency matrix, each representing a polarization state that is orthogonal to the other.

An alternative decomposition is into completely polarized (zero determinant) and unpolarized (scaled identity matrix) components. In either case, the operation of summing the components corresponds to the incoherent superposition of waves from the two components. The latter case gives rise to the concept of the "degree of polarization"; i.e., the fraction of the total intensity contributed by the completely polarized component.

The coherency matrix is not easy to visualize, and it is therefore common to describe incoherent or partially polarized radiation in terms of its total intensity (*I*), (fractional) degree of polarization (*p*), and the shape parameters of the polarization ellipse.

An alternative and mathematically convenient description is given by the Stokes parameters, introduced by George Gabriel Stokes in 1852. The relationship of the Stokes parameters to intensity and polarization ellipse parameters is shown in the equations and figure below.

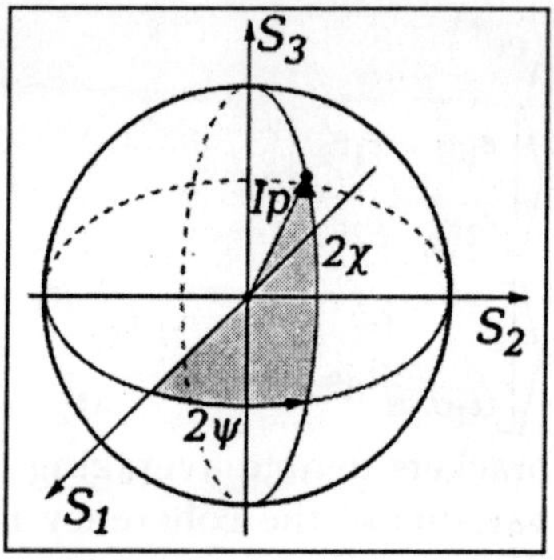

Poincaré sphere diagram

$$S_0 = I$$

$$S_1 = Ip\cos 2\psi \cos 2\chi S_0 = I$$

$$S_2 = Ip\sin 2\psi \cos 2\chi$$

$$S_3 = Ip\sin 2\chi$$

Here *Ip*, 2Ψ and 2χ are the spherical coordinates of the polarization state in the three-dimensional space of the last three Stokes parameters. Note the factors of two before Ψ and χ corresponding respectively to the facts that any polarization ellipse is indistinguishable from one rotated by 180°, or one with the semi-axis lengths swapped accompanied by a 90° rotation. The Stokes parameters are sometimes denoted *I*, *Q*, *U* and *V*.

The Stokes parameters contain all of the information of the coherency matrix, and are related to it linearly by means of the identity matrix plus the three Pauli matrices:

$$\psi = \frac{1}{2}\sum_{j=0}^{3} Sj\sigma_j, \text{ where}$$

$$\sigma_0 = \begin{bmatrix} 1 & 0 \\ 0 & 1 \end{bmatrix} \sigma_1 = \begin{bmatrix} 1 & 0 \\ 0 & -1 \end{bmatrix}$$

$$\sigma_2 = \begin{bmatrix} 0 & 1 \\ 1 & 0 \end{bmatrix} \sigma_3 = \begin{bmatrix} 0 & -i \\ i & 0 \end{bmatrix}$$

Mathematically, the factor of two relating physical angles to their counterparts in Stokes space derives from the use of second-order moments and correlations, and incorporates the loss of information due to absolute phase invariance.

The figure makes use of a convenient representation of the last three Stokes parameters as components in a three-dimensional vector space. This space is closely related to the Poincaré sphere, which is the spherical surface occupied by completely polarized states in the space of the vector

$$\mathrm{u} = \frac{1}{S_0}\begin{bmatrix} S_1 \\ S_2 \\ S_3 \end{bmatrix}.$$

All four Stokes parameters can also be combined into the four-dimensional Stokes vector, which can be interpreted as four-vectors of Minkowski space. In this case, all physically realizable polarization states correspond to time-like, future-directed vectors.

Propagation, Reflection and Scattering

In a vacuum, the components of the electric field propagate at the speed of light, so that the phase of the wave varies in space in time while the polarization state does not. That is,

$$e(z+\Delta z, t+\Delta t) = e(z,t)e^{ikt(c\Delta t - \Delta z)}$$

where k is the wavenumber and positive z is the direction of propagation. The physical electric vector is the real part of the Jones vector. When electromagnetic waves interact with matter, their propagation is altered. If this depends on the polarization states of the waves, then their polarization may also be altered.

In many types of media, electromagnetic waves may be decomposed into two orthogonal components that encounter different propagation effects.

A similar situation occurs in the signal processing paths of detection systems that record the electric field directly. Such effects are most easily characterized in the form of a complex 2×2 transformation matrix called the Jones matrix:

$$e' = Je.$$

In general the Jones matrix of a medium depends on the frequency of the waves.

For propagation effects in two orthogonal modes, the Jones matrix can be written as

$$J = T\begin{bmatrix} g_1 & 0 \\ 0 & g_2 \end{bmatrix} T^{-1},$$

where g_1 and g_2 are complex numbers representing the change in amplitude and phase caused in each of the two propagation modes, and T is a unitary matrix representing a change of basis from these propagation modes to the linear system used for the Jones vectors.

For those media in which the amplitudes are unchanged but a differential phase delay occurs, the Jones matrix is unitary, while those affecting amplitude without phase have Hermitian Jones matrices. In fact, since *any* matrix may be written as the product of unitary and positive Hermitian matrices, any sequence of linear propagation effects, no matter how complex, can be written as the product of these two basic types of transformations.

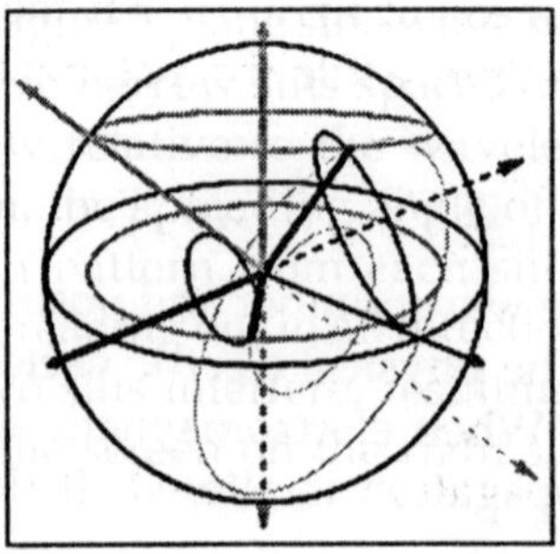

Paths taken by vectors in the Poincaré sphere under birefringence. The propagation modes (rotation axes) are shown with red, blue, and yellow lines, the initial vectors by thick black lines, and the paths they take by colored ellipses (which represent circles in three dimensions).

Media in which the two modes accrue a differential delay are called birefringent. Well known manifestations of this effect appear in optical wave plates/retarders (linear modes) and in Faraday rotation/optical rotation (circular modes). An easily visualized example is one where the propagation modes are linear, and the incoming radiation is linearly polarized at a 45° angle to the modes.

As the phase difference starts to appear, the polarization becomes elliptical, eventually changing to purely circular polarization (90° phase difference), then to elliptical and eventually linear polarization (180° phase) with an azimuth angle perpendicular to the original direction, then through circular again (270° phase), then elliptical with the original azimuth angle, and finally back to the original linearly polarized state (360° phase) where the cycle begins anew.

In general the situation is more complicated and can be characterized as a rotation in the Poincaré sphere about the axis defined by the propagation modes (this is a consequence of the isomorphism of SU(2) with SO(3)). Examples for linear (blue), circular (red), and elliptical (yellow) birefringence are shown in the figure on the left. The total intensity and degree of polarization are unaffected.

If the path length in the birefringent medium is sufficient, plane waves will exit the material with a significantly different propagation direction, due to refraction. For example, this is the case with macroscopic crystals of calcite, which present the viewer with two offset, orthogonally polarized images of whatever is viewed through them. It was this effect that provided the first discovery of polarization, by Erasmus Bartholinus in 1669. In addition, the phase shift, and thus the change in polarization state, is usually frequency dependent, which, in combination with dichroism, often gives rise to bright colors and rainbow-like effects.

Media in which the amplitude of waves propagating in one of the modes is reduced are called dichroic. Devices that block nearly all of the radiation in one mode are known as polarizing filters or simply "polarizers". In terms of the Stokes parameters, the total intensity is reduced while vectors in the Poincaré sphere are "dragged" towards the direction of the favored mode.

Mathematically, under the treatment of the Stokes parameters as a Minkowski 4-vector, the transformation is a scaled Lorentz boost (due to the isomorphism of SL(2,C) and the restricted Lorentz group, SO(3,1)). Just as the Lorentz transformation preserves the proper time, the quantity det $\Psi = S_0^2 - S_1^2 - S_2^2 - S_3^2$ is invariant within a multiplicative scalar constant under Jones matrix transformations (dichroic and/or birefringent).

In birefringent and dichroic media, in addition to writing a Jones matrix for the net effect of passing through a particular path in a given medium, the evolution of the polarization state along that path can be characterized as the (matrix) product of an infinite series of infinitesimal steps, each operating on the state produced by all earlier matrices. In a uniform medium each step is the same, and one may write

$$J = je^{D},$$

where J is an overall (real) gain/loss factor. Here D is a traceless matrix such that αDe gives the derivative of e with respect to z. If D is Hermitian the effect is dichroism, while a unitary matrix models birefringence. The matrix D can be expressed as a linear combination of the Pauli matrices, where real coefficients give Hermitian matrices and imaginary coefficients give unitary matrices. The Jones matrix in each case may therefore be written with the convenient construction

$$J_b = j_b e^{\beta\sigma\cdot\hat{n}} \text{ and } J_r = j_r e^{\phi i\sigma\cdot\hat{m}},$$

where σ is a 3-vector composed of the Pauli matrices (used here as generators for the Lie group SL(2,C)) and n and m are real 3-vectors on the Poincaré sphere corresponding to one of the propagation modes of the medium. The effects in that space correspond to a Lorentz boost of velocity parameter 2β along the given direction, or a rotation of angle 2φ about the given axis.

These transformations may also be written as biquaternions (quaternions with complex elements), where the elements are related to the Jones matrix in the same way that the Stokes parameters are related to the coherency matrix. They may then be applied in pre- and post-multiplication to the quaternion representation of the coherency matrix, with the usual exploitation of the quaternion exponential for performing rotations and boosts taking a form equivalent to the matrix exponential equations above.

In addition to birefringence and dichroism in extended media, polarization effects describable using Jones matrices can also occur at (reflective) interface between two materials of different refractive index. These effects are treated by the Fresnel equations.

Part of the wave is transmitted and part is reflected, with the ratio depending on angle of incidence and the angle of refraction. In addition, if the plane of the reflecting surface is not aligned with the plane of propagation of the wave, the polarization of the two parts is altered.

In general, the Jones matrices of the reflection and transmission are real and diagonal, making the effect similar to that of a simple linear polarizer. For unpolarized light striking a surface at a certain optimum angle of incidence known as Brewster's angle, the reflected wave will be completely *s*-polarized.

Certain effects do not produce linear transformations of the Jones vector, and thus cannot be described with (constant) Jones matrices. For these cases it is usual instead to use a 4×4 matrix that acts upon the Stokes 4-vector.

Such matrices were first used by Paul Soleillet in 1929, although they have come to be known as Mueller matrices. While every Jones matrix has a Mueller matrix, the reverse is not true.

Mueller matrices are frequently used to study the effects of the scattering of waves from complex surfaces or ensembles of particles.

POLARIZATION IN NATURE

POLARIZATION EFFECTS IN EVERYDAY LIFE

Effect of a polarizer on reflection from mud flats. In the picture on the left, the polarizer is rotated to transmit the reflections as well as possible; by rotating the polarizer by 90° (picture on the right) almost all specularly reflected sunlight is blocked.

Fig. The Effects of a Polarizing Filter on the Sky in a Photograph. The Picture on the Right Uses the Filter

Light reflected by shiny transparent materials is partly or fully polarized, except when the light is normal (perpendicular) to the surface. It was through this effect that polarization was first discovered in 1808 by the mathematician Etienne Louis Malus. A polarizing filter, such as a pair of polarizing sunglasses, can be used to observe this effect by rotating the filter while looking through it at the reflection off of a distant horizontal surface.

At certain rotation angles, the reflected light will be reduced or eliminated. Polarizing filters remove light polarized at 90° to the filter's polarization axis. If two polarizers are placed atop one another at 90° angles to one another, there is minimal light transmission.

Polarization by scattering is observed as light passes through the atmosphere. The scattered light produces the brightness and color in clear skies. This partial polarization of scattered light can be used to darken the sky in photographs, increasing the contrast. This effect is easiest to observe at sunset, on the horizon at a 90° angle from the setting sun.

Another easily observed effect is the drastic reduction in brightness of images of the sky and clouds reflected from horizontal surfaces, which is the main reason polarizing filters are often used in sunglasses. Also frequently visible through polarizing sunglasses are rainbow-like patterns caused by color-dependent birefringent effects, for example in toughened glass (e.g., car windows) or items made from transparent plastics.

The role played by polarization in the operation of liquid crystal displays (LCDs) is also frequently apparent to the wearer of polarizing sunglasses, which may reduce the contrast or even make the display unreadable.

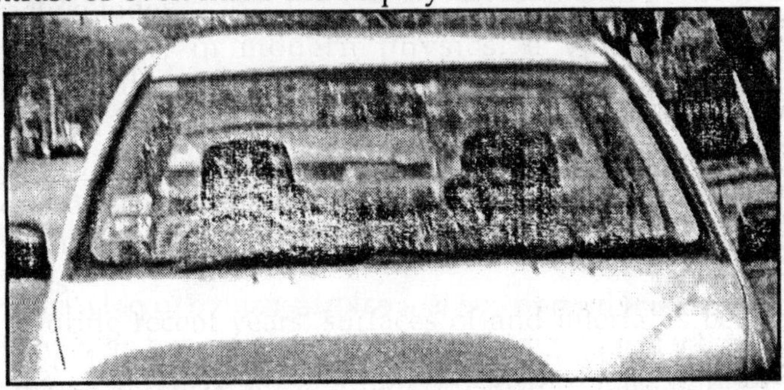

Fig. Polarizing Sunglasses Reveal Stress in Car Window

The photograph on the right was taken through polarizing sunglasses and through the rear window of a car. Light from the sky is reflected by the windshield of the other car at an angle, making it mostly horizontally polarized. The rear window is made of tempered glass. Stress in the glass, left from its heat treatment, causes it to alter the polarization of light passing through it, like a wave plate.

Without this effect, the sunglasses would block the horizontally polarized light reflected from the other car's window. The stress in the rear window, however, changes some of the horizontally polarized light into vertically polarized light that can pass through the glasses. As a result, the regular pattern of the heat treatment becomes visible.

Biology

Many animals are apparently capable of perceiving some of the components of the polarization of light, e.g. linear horizontally-polarized light. This is generally used for navigational purposes, since the linear polarization of sky light is always perpendicular to the direction of the sun. This ability is very common among the insects, including bees, which use this information to orient their communicative dances.

Polarization sensitivity has also been observed in species of octopus, squid, cuttlefish, and mantis shrimp. In the latter case, one species measures all six orthogonal components of polarization, and is believed to have optimal polarization vision. The rapidly changing, vividly colored skin patterns of cuttlefish, used for communication, also incorporate polarization patterns, and mantis shrimp are known to have polarization selective reflective tissue. Sky polarization was thought to be perceived by pigeons, which was assumed to be one of their aids in homing, but research indicates this is a popular myth.

The naked human eye is weakly sensitive to polarization, without the need for intervening filters. Polarized light creates a very faint pattern near the centre of the visual field, called Haidinger's brush. This pattern is very difficult to see, but with practice one can learn to detect polarized light with the naked eye.

Geology

The property of (linear) birefringence is widespread in crystalline minerals, and indeed was pivotal in the initial discovery of polarization. In mineralogy, this property is frequently exploited using polarization microscopes, for the purpose of identifying minerals.

Chemistry

Polarization is principally of importance in chemistry due to the circular dichroism and "optical rotation" (circular birefringence) exhibited by optically active (chiral) molecules. It may be measured using a polarimeter.

Polarization may also refer to the through-bond (inductive or resonant effect) or through-space influence of a nearby functional group on the electronic properties (e.g. dipole moment) of a covalent bond or atom.

Astronomy

In many areas of astronomy, the study of polarized electromagnetic radiation from outer space is of great importance. Although not usually a factor in the thermal radiation of stars, polarization is also present in radiation from coherent astronomical sources (e.g. hydroxyl or methanol masers), and incoherent sources such as the large radio lobes in active galaxies, and pulsar radio radiation (which may, it is speculated, sometimes be coherent), and is also imposed upon starlight by scattering from interstellar dust.

Apart from providing information on sources of radiation and scattering, polarization also probes the interstellar magnetic field via Faraday rotation. The polarization of the cosmic microwave background is being used to study the physics of the very early universe. Synchrotron radiation is inherently polarised.

Technology

Technological applications of polarization are extremely widespread. Perhaps the most commonly encountered examples are liquid crystal displays and polarized sunglasses.

All radio transmitting and receiving antennas are intrinsically polarized, special use of which is made in radar. Most antennas radiate either horizontal, vertical, or circular polarization although elliptical polarization also exists. The electric field or E-plane determines the polarization or orientation of the radio wave. Vertical polarization is most often used when it is desired to radiate a radio signal in all directions such as widely distributed mobile units. AM and FM radio uses vertical polarization. Television uses horizontal polarization. Alternating vertical and horizontal polarization is used on satellite communications (including television satellites), to allow the satellite to carry two separate transmissions on a given frequency, thus doubling the number of customers a single satellite can serve.

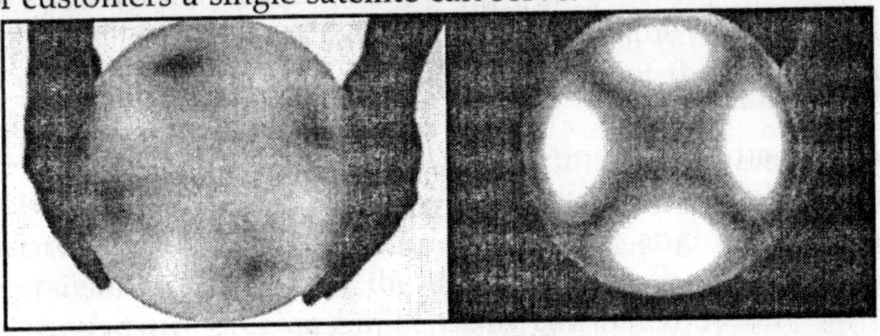

Fig. Strain in Glass

In engineering, the relationship between strain and birefringence motivates the use of polarization in characterizing the distribution of stress and strain in prototypes. Electronically controlled birefringent devices are used in combination with polarizing filters as modulators in fibre optics. Polarizing filters are also used in photography. They can deepen the color of a blue sky and eliminate reflections from windows and standing water.

Sky polarization has been exploited in the "sky compass", which was used in the 1950s when navigating near the poles of the Earth's magnetic field when neither the sun nor stars were visible (e.g. under daytime cloud or twilight). It has been suggested, controversially, that the Vikings exploited a similar device (the "sunstone") in their extensive expeditions across the North Atlantic in the 9th–11th centuries, before the arrival of the magnetic compass in Europe in the 12th century. Related to the sky compass is the "polar clock", invented by Charles Wheatstone in the late 19th century.

Polarization is also used for some 3D movies, in which the images intended for each eye are either projected from two different projectors with orthogonally oriented polarizing filters or from a single projector with time multiplexed polarization (a fast alternating polarization device for successive frames). Filter glasses with similarly oriented polarized filters ensure that each eye receives only the correct image.

Typical stereoscopic projection displays use linear polarization encoding, because it is not very expensive and offers high contrast. In environments where the viewer is moving, such as in simulators, circular polarization is sometimes used. This makes the channel separation insensitive to the viewing orientation. The 3-D effect only works on a silver screen since it maintains polarization, whereas the scattering in a normal projection screen would void the effect.

CAUSES OF POLARIZATION

Polarization is the process that causes neutral parties to take sides in a conflict. It also causes individuals on either side of the conflict to take increasingly extreme positions that are more and more opposed to each other. As parties move toward these opposite "poles," they define themselves in terms of their opposition to a common enemy. Trust and respect diminish, and "distorted perceptions and simplified stereotypes emerge." Parties assume more rigid positions and may refuse to negotiate.

The study of polarization first came to be identified with those realist writers who wrote about the structure of the international system, the impact of military alliances on war and peace, and the balance of power. Writers such as Vasquez, Choucri North, Wallace and Bueno de Mesquita wrote about the effect of polarization on the incidence, severity, and magnitude of great wars and arms races.

Polarization also became the main element in the study of the security dilemma, a situation in which parties feel threatened by an "enemy," so they increase their arms, which causes the other side to feel threatened, therefore increasing their own arms. The dilemma is thus that attempts to bring more security actually bring less. Increased pre-emptive militarization combines with fear, misperceptions, and negative stereotypes to encourage polarization. While polarization can occur in any type of conflict, it has it most damaging effects in large-scale inter-group, public policy, and international conflicts.

Polarization is caused by a number of related psychological, sociological, and political processes. It is closely tied up with escalation in a bi-directional relationship. In other words, escalation causes polarization and vice versa.

Escalation → Polarization → Escalation

As conflict escalates, the emergence of enemy images and stereotypes damages the relationship between adversaries. Important lines of communication and interaction that are normal to peaceful relationships are cut off, and trust diminishes. As parties begin to attribute their grievances to the other side, they often reduce the number of non-conflictual relations and interactions that they have with that party.

Adversaries tend to become increasingly isolated from each other, and any inter-group communication is channeled through more antagonistic lenses. Because parties have fewer ties to individuals from the other group, they may feel freer to employ more severe actions against that group. Group isolation and polarization is further aggravated by the tendency of partisans to try to win bystanders to their side, forcing people to take sides. As more people are drawn into the conflict, that conflict intensifies.

Conversely, escalation seems to increase polarization. Formerly neutral parties are pulled to one side or the other and fewer community members can retain their moderate positions. In part, this is because those involved in the conflict demand that neutral non-participants decide whether they are "with us or against us."

Those who would normally urge moderation and attempt to mediate the conflict are recruited by participants in the controversy, and forced to take sides. It is difficult for community members to remain neutral when people are fighting, damaging each other's property, and injuring each other. In such situations, there is a tendency to cast blame and to side with one party or the other.

Radical positions are further reinforced by group homogeneity and cohesiveness. Kriesberg notes that adversaries with little internal diversity are more prone to escalation. They are more prone to polarization as well. This is because homogeneity makes it less likely that a group will consider alternatives to the severe tactics being advocated or employed by extremists.

As parties assume more radical positions, group members tend to reinforce each other's negative stereotypes and enemy images. Any moderate positions go unheard or their proponents ostracized — or worse — as they are seen as traitors to the cause. As this process continues, parties are often further segregated, and their relationship with outsiders becomes increasingly hostile and competitive.

While some scholars of intergroup conflict regard polarization and escalation as inevitable in interethnic relationships, others see it as the result of social mobilization or manipulation by political elites. Leadership whose legitimacy is threatened, either by the leaders' own actions or by an immediate crisis, can use identity as a 'rallying cry' by calling for mobilization and collective action along nationalistic or ethnic lines. In order to foster homogeneity and build support for their cause, such leaders may portray the adversary as a grave threat to the vital interests and identity of "their people." This furthers both polarization and escalation simultaneously.

Addressing Polarization

Polarization is so much a part of the process of escalation that it is difficult to ascertain if one is the cause of the other. Ikle writes that as soon as two adversaries have initiated violence, their stakes and expectations change, making it impossible to return to a peacetime relationship without first repairing the damage. Escalation has multiple dimensions; it could be a shift or change in the pattern of the violence, but Ikle notes that it also usually prolongs the war by default.

Parties engaged in conflict typically focus on their differences, which can result in pushing the parties toward polar opposite positions. Strategies that encourage parties to consider their common interests can help to mitigate such effects.

Planning and pursuing joint projects, for example, can help parties to focus on commonalities rather than contentious issues. Because fear and distrust often play a central in producing polarization, trust building strategies are also beneficial.

The establishment of personal relationships between adversaries can help to improve their communication, increase their level of mutual understanding, and make them less likely to view each other as evil enemies. Better understanding of a party's true motives can help to reduce anger and hostility and eliminate actions that unwittingly threaten or annoy one's opponent.

If possible, third-party mediators or consultants should help parties to better understand the dynamics of their conflict and to address their negative perceptions and attitudes. This sort of transformative mediation can enhance trust and help parties to refrain from taking hostile actions.[12] If parties can move toward productive negotiation, they have taken the first step towards reconciliation.

ELECTROMAGNETIC FIELD POLARIZATION

The electromagnetic field in the absence of matter, that is, in the "vacuum," is excellently described classically by Maxwell's theory, which is relativistically correct, in terms of the fundamental field vectors E and B. There is an equally successful quantum field theory. In both cases, however, we must introduce sources of the field, and these sources lie in matter, the rigorous description of which is extremely difficult.

In Maxwell's theory, charge is introduced as the source of the electromagnetic field, and is idealized to point and continuous distributions. Matter is separated into conductors and dielectrics depending on whether the fields can penetrate into the matter.

Conductors carry free charges that move to positions of equilibrium in response to the fields, but cannot leave the matter. Dielectrics are presumed to contain bound charges that cannot move freely, but can shift positions when in the presence of the field, which exerts forces upon them, producing polarization, and nonuniform polarization gives rise to new sources of the field. Simple laws are postulated to account for these effects, such as the dielectric constant κ that relates the D and E vectors: $D = \kappa E$ (we'll define these vectors later, but they are probably familiar to us). This theory is called *phenomenological* because all the parameters are empirical, and are not defined in terms of the constitution of the matter.

The problem to be discussed here is the determination of these empirical parameters in terms of the molecular constitution of matter, especially the dielectric constant, historically the first such parameter to be studied, even before the appearance of Maxwellian electrodynamics. This is a huge area of study, and we will only discuss the first step in it, for the simple case of electrostatics, and its extension to electromagnetic waves. The general problem is so complicated that there is no comprehensive, relativistically correct description of the electromagnetic field in matter, although practically a great deal is known, especially in the nonrelativistic case.

The "real" electric field in vacuum is E, giving the force on a point charge q in vacuum: $F = qE$. It is a vector field, obeying the differential laws $\text{div } E = 4\pi\rho$ and $\text{curl } E = 0$, where ρ is the charge density. This leads to the expressions $E = -\text{grad } \varphi$, $\varphi = \int \rho dV/R$, where φ is the scalar potential, and R is the distance from the charge element ρdV to the point P where the potential is desired.

If we know the charge distribution we can, in principle, find the electric field at any point in this way. A point charge q produces a potential $\varphi = q/R$, and so the electric field is $E = -q \text{ grad}(1/R) = (q/R^3)R$, in the direction of R, a vector from the source point to the observation point.

$$\phi(p) = -p \cdot \Delta(1/r) = \frac{p \cdot r}{r^3}$$

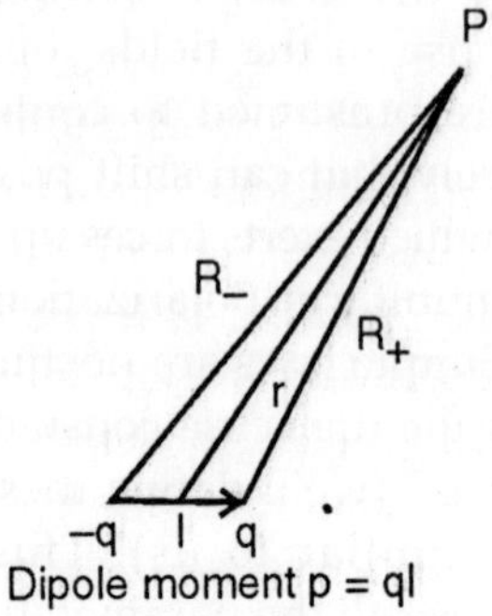

Fig. Dipole Potential

Equal and opposite charges ±q separated by a distance l directed from the negative charge to the positive charge make a finite dipole of dipole moment p = ql, as shown in the figure. The potential produced by this pair of charges is $\varphi = q/R(+) - q/R(-)$, where R(±) are the distances from the charges to the observation point.

The vector R or r points from the dipole to the observation point. We can approximate this difference by differentiating the potential of a point charge: $\varphi = -\text{grad}(1/R) \cdot q\, l = p \cdot R/R^3$. If we idealize the dipole to a point dipole by letting $q \to \infty$ and $l \to 0$ in such a way that p remains constant, then this expression is exact. Such a point dipole is a very good approximation to a finite dipole if we consider observation points at a distance much larger than the actual separation of the charges.

If a charge density ρ and a dipole density P are considered as the sources of the field, then the potential is given by $\varphi = \int [\rho/R + P \cdot \text{grad}(1/R)]dV$. The second term can be cast into a different form by using the vector relation $\text{div}(P/R) = P \cdot \text{grad}(1/R) + \text{div}P/R$ to integrate the dipole part of this expression by parts. Then $\varphi = +\int -\text{div}P dV/R + \int \text{div}(P/R)dV$. The second integral is equal to the surface integral (by the divergence theorem), which vanishes when all space is considered, since P is zero at infinity. Now the potential is $\varphi = \int [\rho - \text{div}P]dV/R$, which shows that -divP is equivalent to charge density. Therefore, we now find that div E = 4πρ - 4π div P, so that div(E + 4πP) = 4πρ. This shows that a new field, D = E + 4πP, can be defined that has divergence only at "free" charge ρ, like the original E without polarization. Polarization provides "bound" charge ρ' = -divP. This new field, called the *displacement*, is found to have useful properties, usually involving different boundary conditions than those applying to E. In free space, there is no polarization, so there D = E.

D satisfies the equation divD = 4πρ. If there is no free charge, ρ = 0, then taking the divergence of D we find divE = -4πdivP = 4πρ', an easy way to remember the charge resulting from nonuniform polarization. At a boundary between two media where there is no free charge on the boundary, the normal component of D is continuous, while the normal component of E has a jump of 4π times the jump in polarization.

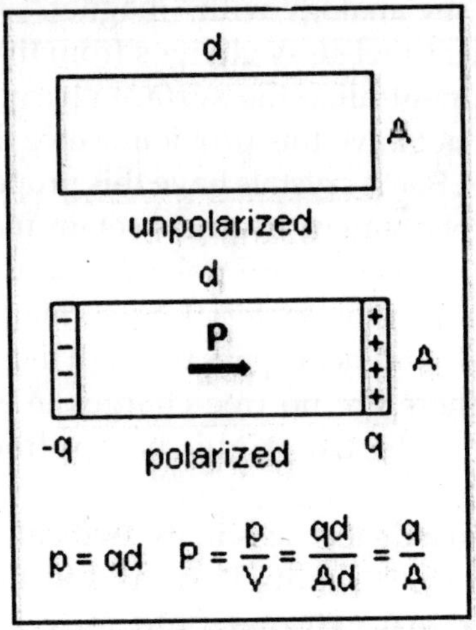

Fig. Polarization

An illustration of these things is shown in the figure at the right. At the top there is a cylinder of length d and end area A, with superimposed equal uniform densities of positive and negative charge, so that there is no external field. This is the case in ordinary matter, where the positive and negative charges cancel, though the total amounts of each is very large, and the matter is said to be *unpolarized*. Now imagine the positive charge moved slightly with respect to the negative charge, so that positive charge of amount q is uncovered at the right, and negative charge of amount -q at the left. The total dipole moment that results is qd.

The polarization P is the dipole moment per unit volume, qd/V = q/A. P is uniform throughout the cylinder. It has a discontinuity in the normal component of -P = -q/A at the right face, which is accompanied by a surface bound charge density of q/A. The discontinuity of +q/A at the left face is accompanied by a bound charge density of -q/A. The normal component of P is equivalent to a surface charge density, $P_n = \sigma'$. At the lateral surfaces, there is a discontinuity of the transverse component of P, but this is accompanied by no surface charge density.

A body such as this with a permanent polarization P is called an *electret*, by analogy with "magnet." Actual electrets, left exposed, will attract stray charges from the surroundings, which will soon neutralize the surface charges.

Magnets do not have this problem, since there are no free magnetic charges. Some crystals have this property of retaining polarization, as permanent magnets retain magnetism. Most materials do not, however, and polarization is produced only by an electric field in the material. Such materials are called *dielectrics*, since electric fields pass through them, in distinction to conductors. There are no free charges in a dielectric that move when an electric field is applied, to a good approximation.

The charge in a dielectric is in its molecules, where strong electric forces hold the light electrons and the heavy nuclei closely together. An external field moves these charges slightly, separating the positive and negative charges to produce polarization. The usual external fields are much smaller than the molecular fields, so to a good approximation the charge separation is proportional to the applied external field.

This is expressed by $P = \chi E$, where the constant χ is the *susceptibility*. The relation between polarization and applied field can be much more complicated than this, even involving the time, but this simplest case covers most applications. Then, $D = E + 4\pi\chi E = \kappa E$, where κ is the *dielectric constant*. It is clear that κ expresses properties of matter in relation to the electric field, and is not fundamental.

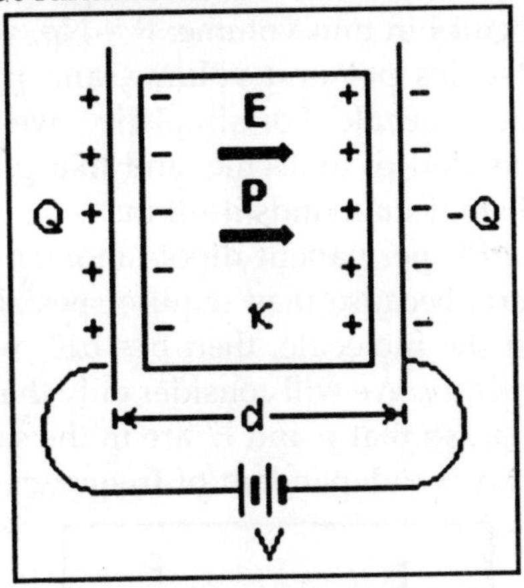

Fig. Capacitor with Dielectric

The capacitor with dielectric shown in the diagram at the left is a familiar example. The space between the plates and the dielectric is exaggerated; they should both be considered a distance d apart, and of area A. The applied voltage V produces an electric field of magnitude $E = V/d$ between the plates, whether the dielectric is present or absent.

This means that the net surface charge density on the plates is $E/4\pi$. If there is no dielectric, this is the actual free charge density on the plates, supplied by the battery. In the presence of the dielectric, $D = \kappa E$ (since all the vectors are parallel, we work with the magnitudes only), and $P = \chi E$. The free charge density on the plates is $D/4\pi$ or $\kappa E/4\pi$, κ times what it was in the absence of the dielectric. The total charge on the plates is $Q = \kappa AV/4\pi d = (\kappa A/4\pi d)V = CV$, where the *capacitance* is $C = \kappa A/4\pi d$.

We can look at this another way. The bound charge density is $\sigma' = P_n = \chi E$, while the net charge density is $\sigma'' = E/4\pi$. Hence, the free charge density on the plates is $\sigma = E/4\pi + \chi E = (1 + 4\pi\chi)E/4\pi = \kappa E/4\pi$. Note that the field used in χE is the actual field in the dielectric, produced by both free and bound charge, not just the field produced by the free charge. The electric field in the narrow region between the plate and the dielectric is larger than E; in fact, it is κE = D.

The polarization P is the sum of the polarizations of the individual molecules in unit volume: P = Np, where N is the number of molecules per unit volume, and p is the dipole moment of each molecule. For simplicity, we consider that there is only one kind of molecule, and that p is induced by the field that the molecule finds itself in.

Molecules with permanent dipole moments, like water, are excluded here, because they require special treatment. If E′ is the field at the molecule, then p = αE′, where α is the *molecular polarizability*. We will consider only the simplest case when α is a scalar, so that p and E′ are in the same direction, and further that α is independent of frequency.

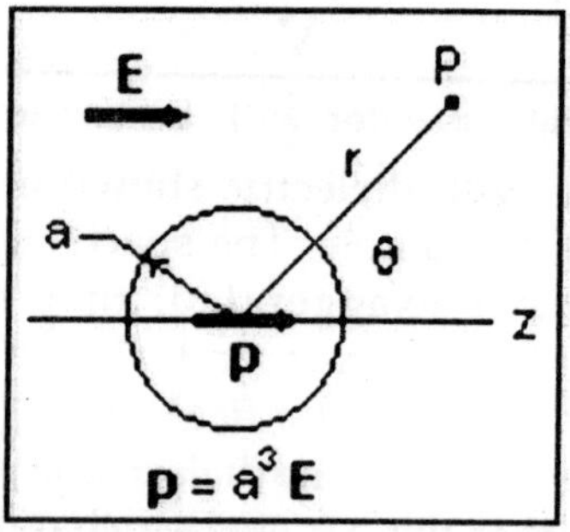

Fig. Cunducting Sphere

The meaning of the molecular polarizability can be made clearer by considering the dipole moment induced in a conducting sphere of radius a by an applied field E. This is a standard potential problem, in which the potential satisfies div grad φ = 0 (this is the Laplacian, del squared), since there is no free charge or polarization in the space where we desire the potential. We know the solutions for this simple geometry, and try the function $\varphi = -Ar\cos\theta + B\cos\theta/r^2$.

If it satisfies the boundary conditions, then it is the solution. First, we see that A = E, so at large distances the potential is -Ez, where z = r cosθ, appropriate for the uniform applied field. Next, φ must be constant (independent of θ) when $r = a$, which means that $-Ea + B/a^2 = 0$, or $B = Ea^3$. Finally, $\varphi = -Ez + p\cos\theta/r^2$, which is just the superimposed field of a uniform applied field and a dipole field from a dipole of magnitude $p = Ea^3$. The polarizability of a conducting sphere is, therefore, a^3. This gives us an idea of what the size of the molecular polarizability will be, about the cube of the molecular dimensions.

Now we wish to relate the molecular polarizability to the susceptibility and dielectric constant, which are experimentally measureable. If we consider a molecule in the dielectric of the capacitor, we might assume that the field E′ experienced by a molecule is the field E. However, this is not a valid assumption. E is the average field in the dielectric, which can be related to the applied voltage, but it does not reflect the possibly large variations in the field produced by the discrete molecules.

The molecule we have under consideration is surrounded by many others which may change the local field. In a gas the molecules are moving at random, and in a crystal they occupy the lattice sites. Liquids are an intermediate case. In both these cases, it can be shown that the average fields produced by the neighbors of the molecule under consideration are zero. This is not always true, but is true for most substances.

We can, therefore, imagine a spherical region surrounding the molecule to be removed from the dielectric without affecting the local field. This region must be spherical to retain the symmetry. Its size is indefinite, but can be chosen large enough that the rest of the dielectric can be considered as a continuum, if the result does not depend on the size of the sphere.

When this region is removed, the polarization now has a discontinuity on the remaining surface of the dielectric, and the resulting bound charge creates a field that represents the net effect of the other molecules of the dielectric on the one in question. This field must be added to the average field to find the effective field E′.

This is a standard potential problem, which we could solve in the same way that we did the problem of the conducting sphere in a uniform field. We assume a dipole field outside the cavity, and a uniform field inside the cavity, and then apply the boundary conditions. However, we will follow a somewhat different method suggested by Born and Wolf (and not too well explained there). Let the potential in the problem of the field produced by the removed sphere be φ', and the potential we are seeking φ.

If we superimpose these solutions, we find the solution for the undisturbed dielectric, $\varphi = 0$. Therefore, $\varphi = -\varphi' = -\int \mathbf{P} \cdot \mathrm{grad}'(1/R)dV'$, where we are integrating over the primed coordinates. Now, $\mathrm{grad}'(1/R) = -\mathrm{grad}(1/R)$ (since the coordinates appear as $x - x'$, etc., in R), and we can take the gradient outside the integral sign to get $\varphi = \mathbf{P} \cdot \int dV'/R = -\mathbf{P} \cdot \mathrm{grad}\ \varphi''$, where φ'' is the potential of a sphere of uniform charge density -1, a problem we can easily solve by direct integration or Gauss's Law.

By Gauss's Law, we find that the radial field inside the sphere is $-(4\pi/3)r$, so that $\varphi = \mathbf{P} \cdot \mathbf{E}'' = -P(4\pi/3)z$, and the field is $-(d/dz)\varphi = 4\pi P/3$, where we have taken the z-direction in the direction of P. Therefore, the effective field is $E' = E + (4\pi/3)P$. The result is independent of the size of the removed sphere, as it should be. Now we have $p = \alpha E'$, or $P = N\alpha(E + 4\pi P/3)$. If we solve for P, then $P = [N\alpha/(1 - 4\pi N\alpha/3)]E$, and we have found the susceptibility in terms of the molecular polarizability.

The dielectric constant $\kappa = 1 + 4\pi\chi = (1 + 8\pi N\alpha/3)/(1 - 4\pi N\alpha/3)$, from which $(4\pi/3)N\alpha = (\kappa - 1)/(\kappa + 2)$. A relation like this was first obtained by Mossotti in 1850, then by Lorenz in 1869, and refined by Clausius in 1879, and which is usually called the Clausius-Mossotti equation.

Polarizable matter was modeled as an assembly of small conducting spheres in the early studies. Maxwell's theory showed that the index of refraction of light, n, was related to κ by $n^2 = \kappa$, so that the formula could be applied to light as well as to static fields.

H. A. Lorentz, and L. V. Lorenz in 1881, derived a similar formula on the basis of the electron theory in which n^2 replaced κ. This formula is called the Lorentz-Lorenz formula.

It was a very important step in relating Maxwell's phenomenological theory to the molecular structure of matter. Lorenz was a Danish physicist with a name strikingly similar to Lorentz's.

We note that the presence of polarization encourages further polarization, cooperatively, since the effective field E' = E + (4π/3)P increases with P. In fact, if the quantity 4πNα/3 approaches 1, κ increases without limit.

The material then "locks in" to a state of permanent polarization (as in an electret). This actually happens for some crystals, called ferroelectrics, and is analogous to permanent magnetic polarization in ferromagnetic substances. For n = 1.5, 4πNα/3 is about 0.3, so significant cooperation is a normal thing.

The quantity $A = (4\pi/3)N_A\alpha$, where N_A is Avogadro's Number, 6.02×10^{23}, is called the *molar refractivity*. The factor (4π/3) is included just to take it out of subsequent relations. Aside from this, it is simply the polarizability per mole. We have $A = (N/N_A)\{(n^2 - 1)/(n^2 + 2)]$, where N is the number of molecules per unit volume.

The factor (N/N_A) can be written M/d, where M is the molecular weight and d is the density, or, for ideal gases, as RT/p. It is simply the volume in cm^3 occupied by a gram-mole of the substance (22,400 cc at STP). For p in atmospheres, R = 82.057 atm-cc/K is convenient.

The molar refractivity remains remarkably constant as the density or pressure is varied, even when there is a change of state. It even works for water (a polar molecule), for which M = 18 g, d = 1.0 g/cc, and n = 1.334. This gives A = 3.71, and the same value is found for water vapour. It holds as well for mixtures, when the separate values of A are weighted by the relative number of molecules.

It is possible to assign atomic refractivities for various atoms, and then add them up to find the molar refractivity. For example, the atomic refractivity of oxygen is 2.01, and that of hydrogen 1.02, so the molar refractivity of water is predicted to be 2.01 + 2 × 1.02 = 4.05, surprisingly close to the actual value of 3.71. The atomic refractivity of carbon is 2.11, sulphur 8.23 and chlorine 5.72, so the molar refractivities of many organic compounds can be estimated.

As a random example, the substance octane, $(CH_3)(CH_2)_6(CH_3)$, was chosen in the Handbook of Chemistry and Physics, where M = 114.23 g, d = 0.7025 g/cc and n = 1.3974 for NaD light at 20° C were found.

These figures give A = 39.19. This reference gives the atomic refractivities as C = 2.591 and H = 1.028. From these we find that 8 x 2.591 + 18 x 1.028 = 39.23, which is remarkably close agreement.

For gases, where $n \approx 1$, $\alpha = (n^2 - 1)/4N$ and $A = (RT/p)(n^2 - 1)/3$, approximately. For air at 0°C and 1 atm, n = 1.0002926. This gives A = 4.37. The relatively large value is due to the N-N triple bond.

Note on Units

I have used cgs (Gaussian) units here, as do the principal references. Although the units used are immaterial, Gaussian units do not involve arbitrary changes of units without physical significance.

Conversion to practical units is easily made when necessary. The Giorgi (MKS) units used in engineering are only the cgs absolute magnetic units with assorted factors of 10, with the confusing unit changes brought about by the constants ε_o and μ_o, which are merely composed of factors of 4π, 10 and the speed of light c.

To work with Gaussian units, a few defining equations such as $E = q/r^2$, $\varphi = q/r$ and $F = qE$ help to keep units straight. Charge q is in statcoulomb, usually called e.s.u., and one e.s.u. is 10/c coulomb. Here, $c = 3 \times 10^{10}$ cm/s (approximately). The charge q is divided by c in all matters concerned with current in the Gaussian system.

The force F is in dyne = gm-cm/s^2 = 10^{-5} N, and energy is in erg = dyne-cm. 1 erg is 10^{-7} J. The unit of φ, the statvolt, or e.s.u., is 1 erg/statcoulomb or 300 $(c/10^7)$ volts. The unit of capacity, the statfarad, is 1 statcoulomb/statvolt, or $10^9/c^2$ F. This makes an e.s.u. of capacity equal to about 1.11 μF, so this unit is a very practical size, unlike the farad. With the help of these factors, all quantities discussed can be converted to practical units.

Our equation $D = E + 4\pi P$ implies that D, E and P all have the same units. This is easy to check, since Coulomb's law gives $[E]$ = statcoulomb/cm^2, which should be the same as dyne/statcoulomb, as $F = qE$ shows. $[P]$ = statcoulomb-cm/cm^3, from its definition, and this likewise is statcoulomb/cm^2.

Finally, the normal component of D is a surface charge density, which again is statcoulomb/cm^2. From $F = qE = q^2/r^2$, we can find the dimensions of statcoulomb in terms of the centimeter (L), gram (M) and second (T). $[q^2]$ = (gm-cm/s^2)(cm^2), or $[q] = M^{1/2}L^{3/2}T^{-1}$. Now the dimensions of any quantity can be found, and conversions to any system of units carried out. Be aware, though, that different definitions in Gaussian and Giorgi units introduce some hazards.

Since D, E and P all have the same units, equations like $P = \chi E$ are easy to interpret, and χ is dimensionless. In magnetism, Giorgi units make H the basic magnetic vector, by analogy with E, but this is a very unfortunate choice, since B and E are actually analogous, and H and D. This has created such great confusion that many workers have reverted to e.m.u. for magnetic problems. There is no substantive difference, of course, just confusion.

No such confusion arises when we write $B = H + 4\pi M$ in Gaussian units, since it is the same as writing H = B - &4πM to define H, by analogy to the definition of D electrically. Now B has all the sources, while those arising from magnetization are subtracted to get H.

In $D = E + 4\pi P$, E has all the sources, and those arising from polarization are subtracted to get D (remember that bound charge is -div P). There is more to be said about magnetism, but this shows the analogy with electricity.

POLARIZATION AND SPIN

A transparent crystalline substance, now known as calcite, was discovered by a naval expedition to Iceland in 1668, and samples of this "Iceland crystal" were examined by the Danish scientist Erasmus Bartholin, who noticed that a double image appeared when objects were viewed through this crystal.

He found that rays of light passing through calcite are split into two refracted rays. Some of the incoming light is always refracted at the normal angle of refraction for the density of the substance and a given angle of incidence, but some of the incoming light is refracted at a different angle.

If the incident ray is perpendicular to the face of the crystal, the ordinary ray undergoes no refraction and passes straight through, just as we would expect, but the extraordinary ray is refracted upon entering the crystal and again upon departing the crystal.

Bartholin noted that the direction in which the extraordinary ray diverges from the perpendicular as it passes into the crystal depends on the orientation of the crystal about the incident axis.

Thus by rotating the crystal about the incident axis, the second image appearing through the crystal revolves around the first image. This phenomenon could have been observed at any time in human history, and might not have been regarded as terribly significant, but by the middle of the 17th century the study of optics had reached a point where the occurrence of two distinct refracted rays was a clear anomaly. Bartholin called this "one of the greatest wonders that nature has produced".

Christiaan Huygens had previously derived the ordinary law of refraction from his wave theory light by assuming that the speed of light in a refracting substance is the same in all directions, i.e., isotropic. When Huygens learned of the double refraction in the Iceland crystal (also known as Iceland spar) he concluded that the crystal must contain two different media interspersed, and that the speed of light is isotropic in one of these media but anisotropic in the other.

Hence he imagined that two distinct wave fronts emanated from each point, one spherical and the other ellipsoidal, and the directions of the two rays were normal to these two wave fronts. He didn't explain why part of the light propagated purely in one of the media, and part of the light purely in the other. Moreover, he discovered another very remarkable phenomena related to this double refraction that was even more difficult to reconcile with his implicitly longitudinal conception of light waves. He found that if a ray of light, after passing through an Iceland crystal, is passed through a second crystal aligned parallel with the first, then all of the ordinary ray passes through the second crystal without refraction, and all of the extraordinary ray is refracted in the second crystal just as it was in the first.

On the other hand, if the secondary crystals are aligned perpendicular to the first, the refracted ray from the first crystal is not refracted at all in the second crystal, whereas the unrefracted ray from the first crystal undergoes refraction in the second. These two cases are depicted in the figures below.

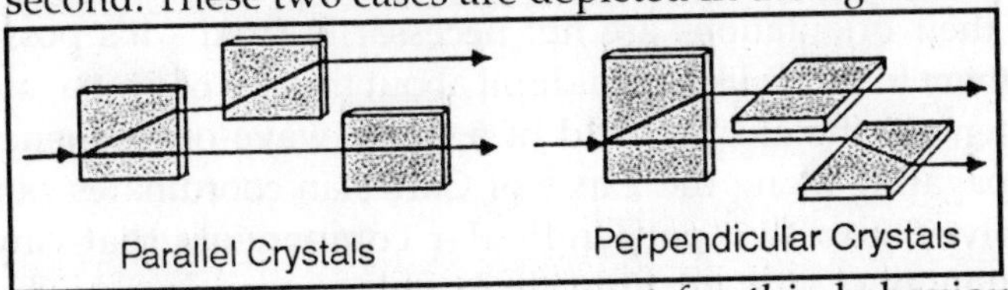

Huygens was unable to account for this behaviour in any way that was consistent with his conception of light as a longitudinal wave with radial symmetry.

Newton considered this phenomena to be inexplicable "if light be nothing other than pression or motion through an aether", and argued that "the unusual refraction is [due to] an original property of the rays", namely, an axial asymmetry or sidedness, which he thought must be regarded as an intrinsic property of individual corpuscles of light.

At the beginning of the 19th century the "sidedness" of Newton of reconciled with the wave concept of Huygens by the idea of light as a transverse (rather than longitudinal) wave.

Later this transverse wave was found to be a feature of the electromagnetic waves predicted by Maxwell's equations, according to which the electric and magnetic fields oscillate transversely in the plane normal to the direction of motion (and perpendicular to each other). Thus an electromagnetic wave "looks" something like this:

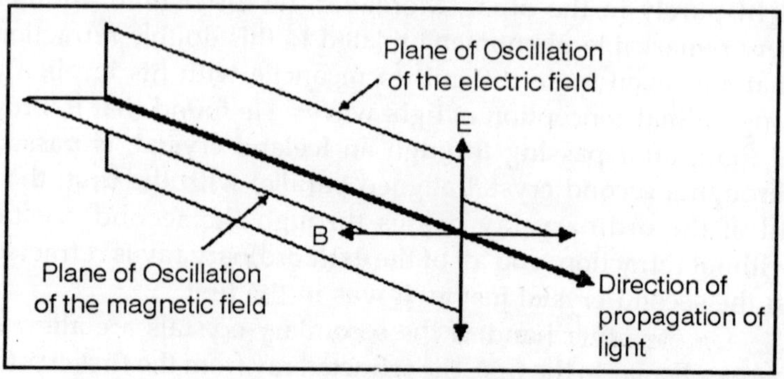

where E signifies the oscillating electric field and B the magnetic field. The wave is said to be polarized in the direction of E. The osculating planes are perpendicular to each other, but their orientations are not necessarily fixed - it's possible for them to rotate like a windmill about the axis of propagation. In general the electric field of a plane wave of frequency ω propagating along the z axis of Cartesian coordinates can be resolved into two perpendicular components that can be written as

$$E_x = C_x \cos(kz - \omega t)$$
$$E_y = C_y \cos(kz - \omega t + \phi)$$

where ϕ is the phase difference between the two components, and C_x and C_y are the constant amplitudes. If the amplitudes of these two components both equal a single constant E_0, and if the phase difference is $-\pi/2$, then remembering the trigonometric identity $\sin(u) = \cos(u - \pi/2)$, we have

$$E_x = E_0 \cos(kz - \omega t)$$
$$E_y = E_y \sin(kz - \omega t)$$

In this case the amplitude of the overall wave is constant, and, as can be seen in the figure below, the electric field vector at constant z rotates (at the angular speed ω) in the clockwise direction as seen by an observer looking back toward the approaching wave.

This is conventionally called right-circularly polarized

light. On the other hand, if the amplitudes are equal but the phase difference is + π/2, then remembering the trigonometric identity sin(u) = -cos(u + π/2), the two components are

$$E_x = E_0 \cos(kz - \omega t)$$
$$E_y = -E_0 \sin(kz - \omega t)$$

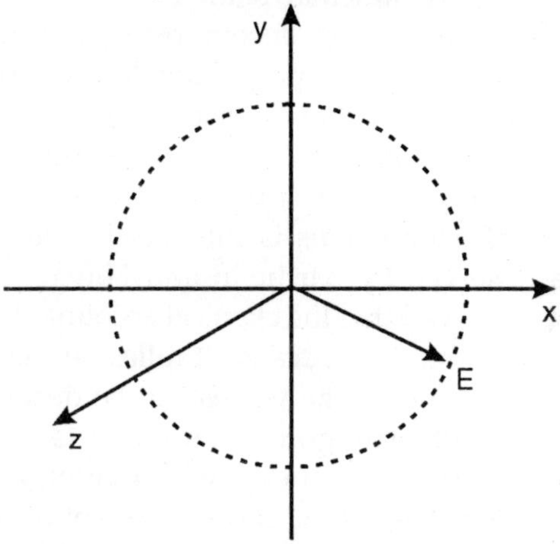

Fig. Right-Circularly Polarized Light

so the direction of the electric field rotates in the counter-clockwise direction. This is called left-circularly polarized light. If we superimpose left and right circularly polarized waves (with the same frequency and phase), The result is simply

$$E_x = 2E_0 \cos(kz - \omega t) \quad E_y = 0$$

which represents a linearly polarized wave, since the electric field oscillates entirely in the xz plane. By combining left and right circularly polarized light in other proportions and with other phase relations, we can also produce what are called elliptically polarized light waves, which are intermediate between the extremes of circularly polarized and linearly polarized light. Conversely, a circularly polarized light wave can be produced by combining two perpendicular linearly polarized waves.

A typical plane wave of ordinary light (such as from the Sun) consists of components with all possible polarizations mixed together, but it can be decomposed into left and right circularly polarized waves, and this can be done relative to any orthogonal set of axes. Calcite crystals (as well as some other substances) have an isotropic index of refraction for light whose electric field oscillates in one particular plane, but an anisotropic index of refraction for light whose electric field oscillates in the perpendicular plane.

Hence it acts as a filter, decomposing the incident wave (normal to the surface) into perpendicular linearly polarized waves aligned with the characteristic axis of the crystal. As the wave enters the crystal, only the component whose electric plane of oscillation encounters anisotropic refractivity is subjected to refraction. This is the classical account of the phenomena observed by Bartholin and Huygens.

It could be argued that the classical account of polarization phenomena is incomplete, because it relies on the assumption that a superposition of plane waves can be decomposed into an arbitrary set of orthogonal components, and that the interactions of those components with matter will yield the same results, regardless of the chosen basis of decomposition.

The difficulty can be seen by considering how a polarizing crystal prevents exactly half of the waves from passing straight through while allowing other half to pass. The incident beam consists of waves whose polarization axes are distributed uniformly in all direction, so one might expect to find that only a very small fraction of the waves would pass through a perfect polarizing substance.

In fact, the fraction of waves from a uniform distribution with polarizations *exactly* aligned with the polarizing axis of a substance should be vanishingly small. Likewise it isn't easy to explain, from the standpoint of classical electrodynamics, why half of the incident wave energy is diverted in one discrete direction, rather than being distributed over a range of refraction angles.

The process seems to be discretely binary, i.e., each bit of incident energy must go in one of just two directions, even though the polarization angles of the incident energy are uniformly distributed over all directions. The precise mechanism for how this comes about requires a detailed understanding of the interactions between matter and electromagnetic radiation - something which classical electrodynamics was never able to provide.

If we discard the extraordinary ray emerging from a calcite polarizing prism, the crystal functions as a filter, producing a beam of a linearly polarized light. The thickness of a polarizing filter isn't crucial (assuming the polarization axis is perfectly uniform throughout the substance), because the first surface effectively "selects" the suitably aligned waves, which then pass freely through the rest of the substance.

The light emerging from the other side is plane-polarized with half the intensity of the incident light. If we pass this polarized beam through another polarizing filter oriented parallel to the first, then all the energy of the incident polarized beam will be passed through the second filter.

On the other hand, if the second filter is oriented perpendicular to the first, none of the polarized beams energy will get through the second filter. For intermediate angles, Etienne Malus (a captain in the army of Napoleon Bonaparte) discovered in 1809 that the intensity of the beam emerging from the second polarizing filter is $I \cos(q)^2$, where I is the intensity of the beam emerging from the first filter and q is the angle between the two filters.

Incidentally, it's possible to convert circularly polarized incident light into plane-polarized light *of the same intensity*. The traditional method is to use a "quarter-wave plate" thickness of a crystal substance such as mica. In this case we're not masking the non-aligned components, but rather introducing a relative phase shift between them so as to force them into alignment. Of course, a particular thickness of plate only "works" this way for a particular frequency.

In 1922 the Dutch physicists Otto Stern and Walther Gerlach made a discovery remarkably similar to that of Erasmus Bartholin, but instead of light rays their discovery involved the trajectories of elementary particles of matter. They passed a beam of particles (atoms of silver) through an oriented magnetic field, and found that the beam split into two beams, with about half the particles in each beam, one deflected up (relative to the direction of the magnetic field) and the other down. This is depicted in the figure below.

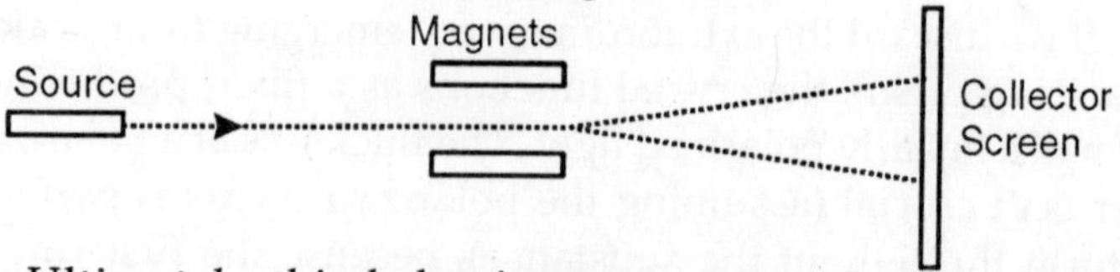

Ultimately this behaviour was recognized as being a consequence of the intrinsic spin of elementary particles. The idea of intrinsic spin was introduced by Uhlenbech and Goudsmit in 1925, and was soon incorporated (albeit in a somewhat ad hoc way) into the formalism of quantum mechanics by postulating that the wave function of a particle can be decomposed into two components, which we might label y_{UP} and y_{DOWN}, relative to any given orientation of the magnetic field.

These components are weighted and the sum of the squares of the weights equals 1. (Of course, the overall state-vector for the particle can be expressed as the Cartesian product of a non-spin vector times the spin vector.) The observable "spin" then corresponds to three operators that are proportional to the Pauli spin matrices:

$$S_x = i\hbar\begin{bmatrix} 0 & 1 \\ 1 & 0 \end{bmatrix} \quad S_y = i\hbar\begin{bmatrix} 0 & -1 \\ i & 0 \end{bmatrix} \quad S_x = i\hbar\begin{bmatrix} 1 & 0 \\ 0 & -1 \end{bmatrix}$$

These operators satisfy the commutation relations

$$\left[S_x, S_y\right] = i\hbar S_z \quad \left[S_x, S_z\right] = i\hbar S_y \quad \left[S_y, S_z\right] = i\hbar S_z$$

as we would expect by the correspondence principle from ordinary (classical) angular momentum. Not surprisingly, this non-commutation is closely related to the non-commutation of ordinary spatial rotations of a classical particle, in the sense that they're both related to the cross-product of orthogonal vectors. Given an orthogonal coordinate system [x,y,z] the angular momentum of a classical particle with momentum $[p_x, p_y, p_z]$ is (in component form)

$$H_x = yp_z - zp_y$$

$$H_y = zp_x - xp_z$$

$$H_z = xp_y - yp_x$$

Guided by the correspondence principle, we replace the classical components p_x, p_y, p_z with their quantum mechanical equivalents, the differential operators

$$-i\hbar\frac{\partial}{\partial x} \quad -i\hbar\frac{\partial}{\partial y} \quad -i\hbar\frac{\partial}{\partial z},$$

leading to the S operators. Although this works, it is not entirely satisfactory, first because it is ad hoc, and second because it is not relativistic. Both of these shortcomings were eliminated by Dirac in 1928 when he developed the first relativistic equation for an elementary massive particle.

The Dirac equation is one of the greatest examples of the heuristic power of the principle of relativity, leading not only to an understanding of the necessity of intrinsic spin, but also to the prediction of anti-matter, and ultimately to quantum field theory. The invariant spacetime interval along the path of a timelike particle of mass m in special relativity is

$$(d\tau)^2 = (dt)^2 - (dx)^2 - (dy)^2 - (dz)^2$$

and if we multiply through by $m^2/(dt)^2$ and make the identifications $E = m(dt/d\tau)$, $p_x = m(dx/d\tau)$, etc., this gives

$$m^2 = E^2 - px^2 - py^2 - pz^2$$

Also, if we postulate that the particle is described by a wave function y(t,x,y,z) we can differentiate with respect to t to give

$$\frac{d\psi}{d\tau} = \frac{\partial\psi}{\partial t}\frac{dt}{d\tau} + \frac{\partial\psi}{\partial x}\frac{dx}{\partial\tau} + \frac{\partial\psi}{\partial y}\frac{dy}{\partial\tau} + \frac{\partial\psi}{\partial z}\frac{dz}{\partial\tau}$$

multiplying through by m and making the identifications for E, p_x, p_y, p_z, we get

$$m\frac{d\psi}{d\tau} = E\frac{\partial\psi}{\partial t} + p_x\frac{\partial\psi}{\partial x} + p_y\frac{\partial\psi}{\partial y} + p_z\frac{\partial\psi}{\partial z}$$

This relation would be equivalent to equation if we put

$$\frac{d\psi}{d\tau} = \frac{m}{i\hbar}\psi \qquad \frac{\partial\psi}{\partial t} = \frac{E}{i\hbar}\psi \qquad \frac{\partial\psi}{\partial x} = -\frac{p_x}{i\hbar}\psi$$

$$\frac{\partial\psi}{\partial y} = -\frac{p_y}{i\hbar}\psi \qquad \frac{\partial\psi}{\partial z} = -\frac{p_z}{i\hbar}\psi$$

where $\hbar$ is a constant. This suggests the operator correspondences

$$i\hbar\frac{d}{d\tau}=m \qquad -i\hbar\frac{\partial}{\partial t}=E \qquad -i\hbar\frac{\partial}{\partial x}=p_x$$

$$-i\hbar\frac{\partial}{\partial y}=p_y \qquad -i\hbar\frac{\partial}{\partial z}=p_z$$

on the basis of which equation can be re-written as

$$\left(\frac{\partial^2}{\partial t^2}-\frac{\partial^2}{\partial x^2}-\frac{\partial^2}{\partial y^2}-\frac{\partial^2}{\partial z^2}+\frac{m^2}{\hbar^2}\right)\psi=0$$

which, if we identify $\hbar$ with the reduced Planck constant $h/(2\pi)$, is the Klein-Gordon wave equation. Unfortunately the solutions of this equation do not give positive-definite probabilities, so it was ruled out as a possible quantum mechanical wave function for a massive particle. Schrodinger's provisional alternative was to base his wave mechanics on the non-relativistic version of equation, which is simply $E = p^2/(2m)$. This led to the familiar Schrodinger equation, whose solutions do give positive-definite probabilities, and which was highly successful in non-relativistic contexts. Still, Dirac was dissatisfied, and sought a relativistic wave equation with positive-definite probabilities.

Dirac's solution was to factor the quadratic equation into linear factors, and take one of those factors as the basis of the quantum mechanical wave equation. Of course, equation doesn't factor if we restrict ourselves to the set of real numbers, but it can be factored in different classes of mathematical entities, just as $x^2 + 1$ can't be factored if we are restricted to real numbers, but it factors as $(x + i)(x - i)$ if we allow imaginary as well as real units. In order to factor equation, Dirac postulated a set of basis variables γ_0, γ_1, γ_2, and γ_3 (not necessarily commuting) such that

$$E^2 - px^2 - py^2 - pz^2 - m^2$$

$$=\left(\gamma_0 E+\gamma_1 p_x+\gamma_2 p_y+\gamma_3 p_z-m\right)\left(\gamma_0 E+\gamma_1 p_x+\gamma_2 p_y+\gamma_3 p_z+m\right)$$

Expanding the product and collecting terms, we find that this is a valid equality if and only if the four variables γ_j satisfy the relations

$$\gamma_0^2=1 \quad \gamma_1^2=\gamma_2^2=\gamma_3^2=-1 \qquad \gamma_1\gamma_j=-\gamma_j\gamma_i$$

for all i, j = 0,1,2,3 with i ≠ j. These four quantities, along with unity, form the basis of what is called a Clifford algebra. The natural representation of these "quantities" is as 4 × 4 matrices. Equation is solved provided either of the factors equals zero. Setting the first factor equal to zero and making the operator substitutions for energy and momentum, we arrive at Dirac's equation

$$\left[\gamma_0 \frac{\partial}{\partial t} - \gamma_1 \frac{\partial}{\partial x} - \gamma_2 \frac{\partial}{\partial y} - \gamma_3 \frac{\partial}{\partial z} - I\frac{m}{i\hbar}\right]\psi = 0$$

Since the operator is four-dimensional, the wave function must be a vector with four components, i.e., we have

$$\psi = \begin{bmatrix} \psi_1 \\ \psi_2 \\ \psi_3 \\ \psi_4 \end{bmatrix}$$

The four components encode the different possible intrinsic spin states of the particle, subsuming the earlier ad hoc two-dimensional state vector. The appearance of four components instead of just two is due to the fact that these state vectors also encompass negative energy states as well as positive energy states. This was inevitable, considering that the relativistic equation is satisfied equally well by –E as well as +E. It may be surprising at first that equation, which is linear, is nevertheless covariant under Lorentz transformations. The covariance certainly isn't obvious, and it is achieved only by stipulating that the components of ψ transform not as an ordinary four-vector, but as two spinors.

Thus the requirement for Lorentz covariance leads directly to the existence of intrinsic spin for any massive particle described by Dirac's equation, including electrons. (Such particles are said to possess "spin 1/2", since it can be shown that the angular momentum represented by their intrinsic spin is $+\hbar/2$.)

In addition, when the interaction with an electromagnetic field is included in the Dirac equation, the requirement for Lorentz covariance leads to the existence of anti-particles. The positron, which is the anti-particle of the electron, was predicted by Dirac around 1929, and found experimentally just two years later.

Fundamentally, the Dirac equation introduced, for the first time, the idea that any relativistic treatment of one particle must inevitably involve consideration of other particles, and from this emerged the concept of second quantization and quantum field theory. In effect, quantum field theory requires us to consider not just the field of a single identified particle, but the field of all such fields. One outcome of quantum field theory was a quantization of the electromagnetic field, the necessity of which had been pointed out by Einstein as early as 1905. On an elementary level, Maxwell's equations are inadequate to describe the phenomena of radiation.

The quantum of electromagnetic radiation is called the photon, which behaves in some ways like an elementary particle, although it is massless, and therefore always propagates at the speed of light. Hence the "spin axis" of a photon is always parallel to its direction of motion, pointing either forward or backward.

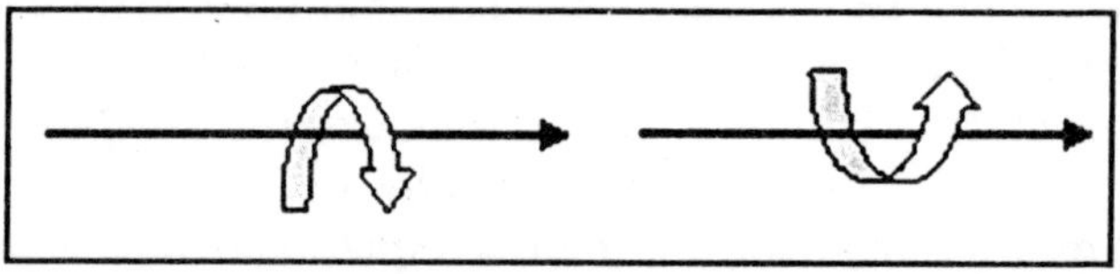

These two states correspond to left-handed and right-handed photons. Whenever a photon is absorbed by an object, an angular momentum of either $+\hbar$ or $-\hbar$ is imparted to the object. Each photon is characterized not only by its energy (frequency) and its phase, but also by it's propensity to exhibit each of the two possible states of spin when it interacts with an object. A beam of light, consisting of a large number of photons, is characterized by the energies, phase relations, and spin propensities of its constituent photons. This could be said to vindicate Newton's belief that rays of light posses a previously unknown "original property" that affects how they are refracted. In classical electromagnetic theory, the plane of oscillation of the electric field of circularly polarized light rotates about the axis of propagation (in one direction or the other). When such light impinges on a surface, it imparts angular momentum due to the rotation of the electric field. In quantum theory this corresponds to a stream of photons with a high propensity for being right-handed (or for being left-handed), so that each photon contributes to the overall angular momentum imparted to the absorbing object.

On the other hand, linearly polarized light (in classical electrodynamics) does not impart any angular momentum to the absorbing object. This is represented in quantum theory by a stream of photons, each with equal propensity to exhibit right-handed or left-handed spin. Each individual interaction, i.e., each absorption of a photon, imparts either $+\hbar$ or $-\hbar$ to the absorbing object, so if the intensity of a linearly polarized beam of light is lowered to the point that only one photon is transmitted at a time, it will appear to be circularly polarized (either left or right) for each photon, which of course is not predicted by classical theory. (In a sense, Maxwell's equations can be regarded as a crude form of the Schrφdinger equation for light, but it obviously does not represent all the quantum mechanical effects.) However, for a large stream of such photons, the net angular momentum is essentially zero, because half of the photons interact in the right-handed sense and half in the left-handed sense.

This corresponds (loosely) to the fact in classical theory that linearly polarized light can be regarded as a superposition of left and right circularly polarized light. Incidentally, most people have personal "hands on" knowledge of polarized electromagnetic waves without even realizing it. The waves broadcast by a radio or television tower are naturally polarized, and if we 've ever adjusted the orientation of "rabbit ears" and found that our reception is better at some orientations than at others, for a particular station, we 've demonstrated the effects of electromagnetic wave polarization.

The behaviour of intrinsic spin of elementary particles can be used to illustrate some important features of quantum mechanics - features which Einstein famously referred to as "spooky action at a distance".

Chapter 7

Scattering Cross Sections

A scattering cross section into which particles are scattered during the collision with the force field of another particle was first developed by Ernest Rutherford. The idea behind the scattering cross section is that a mathematical relationship can be derived which will relate the probability that particles will be scattered into a given angle.

It is important to understand the derivation of the scattering cross section, because much of its importance comes from the equations that take form during the derivation. Once the equations governing the scattering cross section have been formulated, it is then necessary to understand Rutherford's experiment and equations. The usefulness of scattering cross sections in atomic and subatomic physics can be realized from the Rutherford model.

By the start of the 20th century, Sir Joseph John Thomson had developed the "plum pudding" model of the atom. Thomson suggested that an atom was comprised of a sphere of positive charge into which electrons were embedded, like raisins in a loaf of raisin bread or plums in a bowl of plum pudding.

In 1909 British scientist Ernest Rutherford along with Hans Geiger and Ernest Marsden, performed experiments to test the Thomson model of the atom. Rutherford's experiment included firing a beam of alpha particles at a sheet of gold foil. In order for Thomson's model to hold true the alpha particles could not be deflected by more than a fraction of a degree. However, Rutherford saw large deflection angles for the alpha particles. According to Rutherford, this proved that the atoms contained particles that were at least as heavy as the alpha particles themselves.

With the results of his experiments, Rutherford proposed a model of the atom which included a dense nucleus. Through the idea of a scattering cross section for a particle encountering the force field of another particle, Rutherford was able to develop his scattering formula for the Coulomb potential. This formula not only predicts the angle at which charged particles will be deflected, but also provides a means for estimating nuclear radii.

SCATTERING CROSS SECTIONS IN THE CENTRE OF MASS

To understand the ideas behind scattering cross sections, it is first necessary to understand the physics behind a particle moving in a central-force field.

It is known and can be derived from Langrangian dynamics applied to a particle moving in a central force field that the angular momentum of the particle is conserved:

$$l = \mu r^2 \dot{\theta} = \text{constant or } \dot{\theta} = \frac{l}{\mu r^2}$$

where μ is the reduced mass of the system of particles.

In the absence of dissipative forces the total energy of the system must also be conserved:

$$E = T + U = \text{constant}$$

where

$$E = \frac{1}{2}\mu\left(\dot{r}^2 + r^2\dot{\theta}^2\right) + U(r)$$

By substituting the value of the angular velocity found from the conservation of angular momentum, the energy takes the following form:

$$E = \frac{1}{2}\mu\dot{r}^2 + \frac{1}{2}\frac{l^2}{\mu r^2} + U(r)$$

This equation may be solved for the radial velocity:

$$\dot{r} \pm \sqrt{\frac{2}{\mu}(E - U) - \frac{l^2}{\mu^2 r^2}}$$

From the chain rule:

$$d\theta = \frac{d\theta}{dt}\frac{dt}{dr}dr = \frac{\dot{\theta}}{\dot{r}}dr$$

It follows that from Equations:

$$\theta(r) = \int \frac{\pm\left(\frac{l}{r^2}\right)dr}{\sqrt{2\mu\left(E-U-\frac{l^2}{2\mu r^2}\right)}}$$

From Equation, the radial velocity vanishes at the roots of the radical. This indicates at these points the particle is experiencing a turning point. These two roots represent the maximum and minimum radii achieved by the orbit. The angular change in the orbit is twice that which would result from travel from the minimum to the maximum:

$$\Delta\theta = 2\int_{r\min}^{r\max} \frac{\left(l/r^2\right)dr}{\sqrt{2\mu\left(E-U-\frac{l^2}{2\mu r^2}\right)}}$$

This final equation will be used in the analysis of the scattering angle experienced by a particle as it approaches the central force field.

The differential scattering cross section is defined as the number of interactions per target particle that lead to scattering into an element of the solid angle at a given angle divided by the number of incident particles per unit area. This figure illustrates the interaction in a centre of mass reference frame:

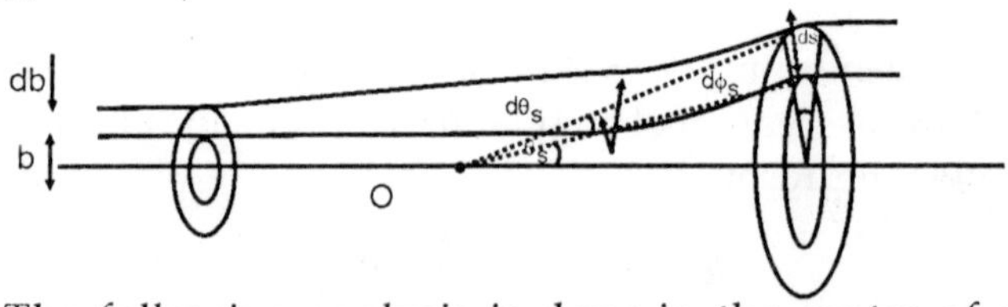

The following analysis is done in the centre of mass reference frame. An incident particle with impact parameter b is deflected through an angle θ_s into a detector at a great distance L from the scattering centre O. The differential surface area of the detector is represented by ds.

The solid angle through which the particles are deflected in Figure will be represented $d\Omega_s$ as and may be obtained through the following relationship. It is important to look Figure to understand the relationship between the differential surface area and the deflection angle.

Since the force is considered to be a central force, it must possess axial symmetry eliminating the azimuthal angle from the previous relationship:

$$d\Omega_s = 2\pi \sin\theta_2 d\theta_2$$

This defines the solid angle for scattering in the differential cross section analysis.

The probability of scattering into the solid angle $d\Omega_s$ for a unit of area of the incident beam is:

$$\sigma(\theta)\,d\Omega_s = \frac{dN}{I} \longrightarrow \qquad dN = I\sigma(\theta)\,d\Omega_s$$

Where $\sigma(\theta)$ is the differential cross section, dN is the number of particles scattered into $d\Omega_s$ per unit time, and I is the intensity of the beam (the number of particles per unit time passing through a unit area normal to the beam.)

From Figure the incident particles pass through a ring of area:

$$dA_b = bd\phi db$$

The number of particles passing through this ring must be equal to the number of particles passing through the differential area of the detector, dN.

$$I \cdot dA_b = Ibd\phi db = I \cdot 2\pi bdb$$

$$I \cdot 2\pi bdb = I\sigma(\theta) \cdot 2\pi \sin\theta_s d\theta_s$$

This gives the equation for the differential cross section:

$$\sigma(\theta) = -\frac{b}{\sin\theta_s} \cdot \frac{db}{d\theta_s}$$

where the equation is negative because an increase db in the impact parameter means less force is exerted on the particle resulting in a decrease in the scattering angle.

Now a closer look at the deflection of the particle is required to determine the dependence of the scattering angle on the impact parameter. The following figure illustrates the scattering of the particle just at the scattering centre.

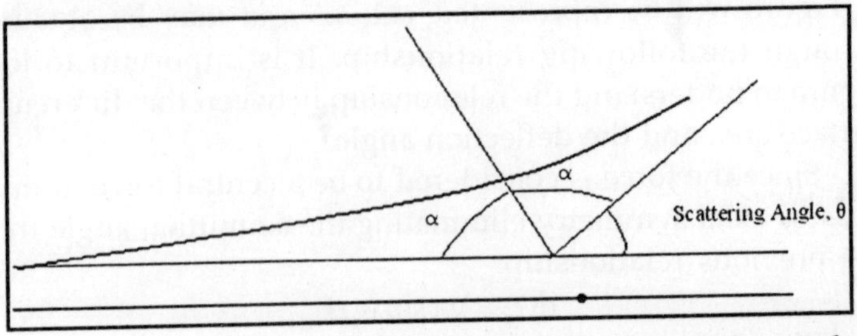

From this figure it can be seen that the scattering angle is equal to 180° minus twice the angle $\alpha\,(\theta = \pi - 2\alpha)$. Therefore if the angle α can be solved for, then the angle θ will be known. From Equation 7 the angular change of the orbit in the central field problem is equal to the angle α in Figure. The integral will be from r_{min} to $r_{max}=\infty$ since the orbit is of a hyperbolic form. The value r_{min} is as stated previously the root of the radical in Equation. This corresponds to the turning point or distance of closest approach for the particle. Therefore, an equation can be derived that gives " as a function of the impact parameter only:

$$a = \int_{r\min}^{\infty} \frac{\left(l/r^2\right)dr}{\sqrt{2\mu(E-U)-\dfrac{l^2}{r^2}}}$$

Now by making use of the equations that relate the angular momentum and kinetic energy of a particle to the impact parameter b $\left(T_0 = \frac{1}{2}\mu u_1^2 \text{ and } l = b\sqrt{2\mu T_0}\right)$, an equation is derived that explicitly relates the impact parameter and scattering angle when the potential and kinetic energies are known or can be solved for.

$$a = \int_{r\min}^{\infty} \frac{\left(b/r^2\right)dr}{\sqrt{1-\left(b^2/r^2\right)-\left(U/T_0\right)}}$$

where $E=T_0$ because the total energy must equal the

kinetic energy at $r = \infty$ where the potential energy is equal to zero.

This equation along with the others derived in this analysis can be used to relate the impact parameter and scattering angle of a two particle system undergoing a central force field collision in the centre of mass reference frame.

CONVERSION FROM CENTRE OF MASS TO LABORATORY REFERENCE FRAME

Since laboratory measurements are made in the reference frame of the lab itself and not the centre of mass reference frame, it is now necessary to derive the equations that will allow for conversion from one reference frame to the other. The following figure will be helpful in identifying the parameters that connect the two reference frames:

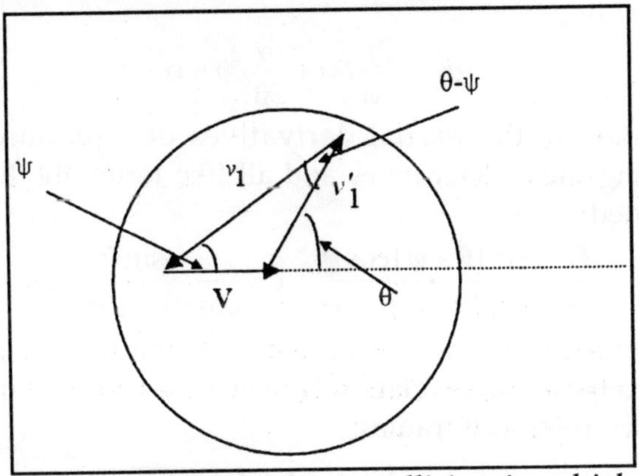

Figure represents an elastic collision in which there is one trajectory. The lower case v's represent the velocity of the particle in two different reference systems before the collision and V represents the resultant velocity. The centre of mass velocity is represented by a prime variable. The total number of particles scattered into the solid angle must be the same in both reference frames. Now using prime variables to represent all variables in the centre of mass coordinate system:

$$I = \sigma(\theta)d\Omega' = I\sigma(\psi)d\Omega$$

$$\sigma(\theta)\cdot 2\pi\sin\theta d\theta = \sigma(\psi)\cdot 2\pi\sin\psi d\psi$$

In order to obtain the proper conversion Equation 16 must be solved for $\sigma(\psi)$:

Now from the Law of Sines and Figure:

$$\frac{\sin(\theta-\psi)}{V}=\frac{\sin\psi}{v_1'}$$

From conservation laws the following relate the mass and velocity of the particles in the two reference frames:

$$V=\frac{m_1u_1}{m_1+m_2} \text{ and } v_1'=V=\frac{m_2u_1}{m_1+m_2}$$

By combining Equations 18 and 19:

$$\frac{\sin(\theta-\psi)}{\sin\psi}=\frac{V}{v_1'}=\frac{m_1}{m_2}=\chi=\text{constant}$$

The value of χ is constant meaning $d\chi = 0$. From the chain rule:

$$d\chi=\frac{\partial\chi}{\partial\psi}\partial\psi+\frac{\partial\chi}{\partial\theta}\partial\theta=0$$

By taking the partial derivatives of Equation 20 and combining these derivatives and all like terms the following is obtained:

$$\frac{d\theta}{d\psi}=\frac{\sin(\theta-\psi)\cos\psi}{\cos(\theta-\psi)\sin\psi}+1=\frac{\sin\theta}{\cos(\theta-\psi)\sin\psi}$$

After some mathematical manipulation the equation can be solved for a usable relation between the centre of mass and laboratory reference frames:

$$\sigma(\psi)=\sigma(\theta)\cdot\frac{\left[x\cos\psi+\sqrt{1-x^2\sin\psi}\right]^2}{\sqrt{1-x^2\sin^2\psi}}$$

and

$$\theta=\sin^{-1}(x\sin\psi)+\psi$$

THE RUTHERFORD SCATTERING EQUATION AND THE COULOMB POTENTIAL

While much of the discussion thus far has been mathematical, it is important to understand these mathematics to use the idea of scattering cross sections in a practical manner. One of the most notable uses of this idea comes from the Rutherford scattering equation.

By using the method outlined in the mathematics portion of this paper, it is fairly simple to derive an equation for the scattering cross section of a particle experiencing an interaction with a nucleus. The equation for this cross section is:

$$\sigma(\theta) = \frac{k^2}{(4T_o)^2} \cdot \frac{1}{\sin^4(\theta/2)}$$

where $k = \frac{q_1 q_2}{4\pi\varepsilon_o}$ and ε_o is the permitivity of free space.

This equation can be rewritten in a more functional form. In this form parameters are either a property of the material or can be measured experimentally. The following represents the number of particles per unit area striking the detector:

$$N(\theta) = \frac{N_i n L Z^2 k^2 e^4}{4r^2 T^2 \sin^4(\theta/2)}$$

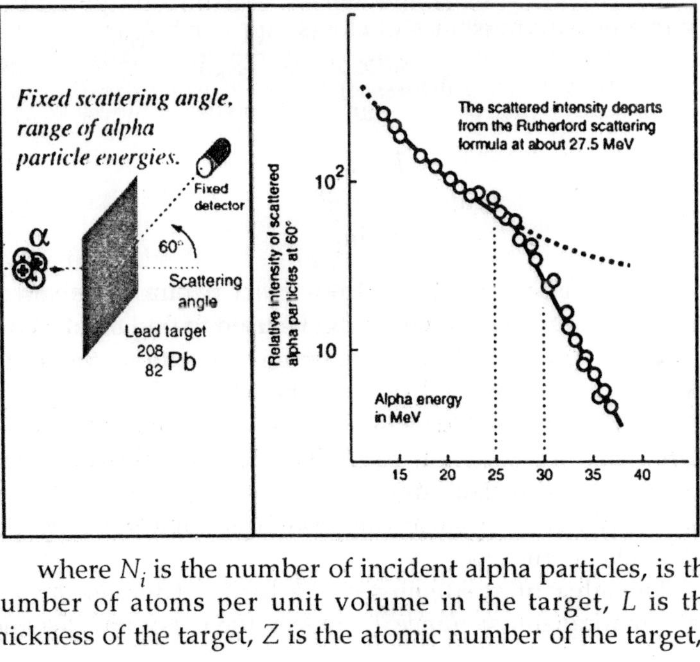

where N_i is the number of incident alpha particles, is the number of atoms per unit volume in the target, L is the thickness of the target, Z is the atomic number of the target, e

is the electron charge, k is Coulomb's constant, r is the target-to-detector distance, T is the kinetic energy of the alpha particles, and θ is the scattering angle.

While this formula can prove to be a valuable tool in predicting scattering of particles entering a Coulomb potential, its real experimental value comes from a particle's departure from this formula. The following depicts the departure of the Rutherford formula for scattering ^{208}Pb:

This data is important because it allows for the calculation of the radius of the lead nucleus. With alpha particles at high enough energies, the particles gets close enough to the nucleus to become effected by the nuclear strong forces and cause the departure. The energy of the particles that begin to depart from the Rutherford equation can then be used to calculate the impact parameter and distance of closest approach to the nucleus. The value of the distance of closest approach can then be used to estimate the nuclear radius.

The following equations were derived from the scattering cross section equations and can be used to find the impact paramater b and distance of closest approach r_{min}:

$$b = \frac{ke^2 Z_1 Z_2}{mv^2}\sqrt{\frac{1+\cos\theta}{1-\cos\theta}}$$

$$r\min = \frac{b\cos\left(\frac{\theta}{2}\right)}{1-\sin\left(\frac{\theta}{2}\right)}$$

It can be seen that once the impact parameter is solved for in Equation, then Equation can be used to find the distance of closest approach. The value of the velocity of the alpha particles must be solved for from the energy that caused the departure from Rutherford's equation, because this is the energy that allowed the particles to punch through close enough to the nucleus to cause the departure. This means that the distance of closest approach is approximately the radius of the nucleus.

The idea of scattering cross sections seems mostly mathematical at first glance. However, the use of the idea and the equations governing scattering cross sections can be used to solve for other parameters that cannot be measured directly in a laboratory.

Rutherford's experiment completely revolutionized ideas about the structure of the atom. Rutherford's discovery of the nucleus laid the foundation for the modern atomic model. While this discovery had a great impact on scientific history, the equations developed by Rutherford continue to provide valuable tools for laboratory measurement.

Chapter 8

Waves in Two and Three Dimensions

The extension of the sine wave to higher dimensions is the plane wave. Wave packets in two and three dimensions arise when plane waves moving in different directions are superimposed.

Diffraction results from the disruption of a wave which is impingent upon an object. Those parts of the wave front hitting the object are scattered, modified, or destroyed. The resulting *diffraction pattern* comes from the subsequent interference of the various pieces of the modified wave. A knowledge of diffraction is necessary to understand the behaviour and limitations of optical instruments such as telescopes.

Diffraction and interference in two and three dimensions can be manipulated to produce useful devices such as the diffraction grating.

VECTORS

Before we can proceed further we need to explore the idea of a vector. A vector is a quantity which expresses both magnitude and direction. Graphically we represent a vector as an arrow. In typeset notation a vector is represented by a boldface character, while in handwriting an arrow is drawn over the character representing the vector.

Figure shows some examples of displacement vectors, i. e., vectors which represent the displacement of one object from another, and introduces the idea of vector addition.

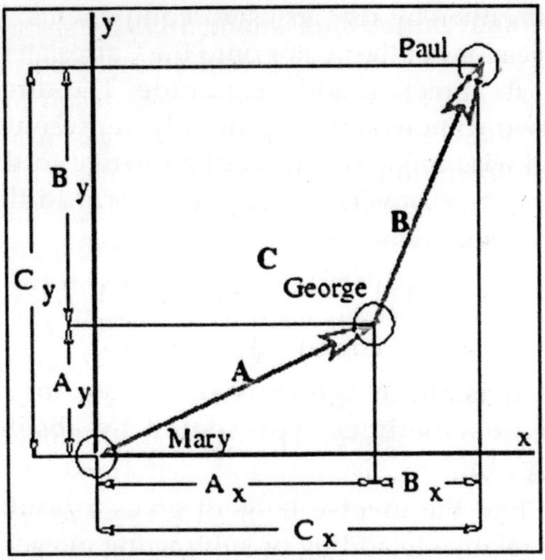

Fig. Displacement vectors in a plane. A Vector represents the displacement of George from Mary, while vector B represents the displacement of Paul from George. Vector represents the displacement of Paul from Mary and C = A + B. The quantities, A_x, A_x, etc., represent the Cartesian components of the vectors.

The tail of vector B is collocated with the head of vector A, and the vector which stretches from the tail of A to the head of B is the sum of A and B, called C in figure.

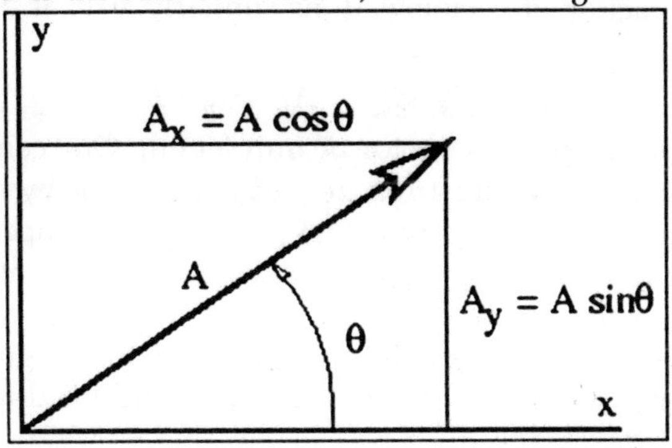

Fig. Definition Sketch for the Angle Representing the Orientation of a Two Dimensional Vector

The quantitiesA_x, A_y, etc., represent the Cartesian components of the vectors in figure. A vector can be represented either by its Cartesian components, which are just the projections of the vector onto the Cartesian coordinate axes, or by its direction and magnitude. The direction of a vector in two dimensions is generally represented by the counterclockwise angle of the vector relative to the axis, as shown in figure. Conversion from one form to the other is given by the equations

$$A = \left(A_x^2 + A_y^2\right)^{1/2} \qquad \theta = \tan^{-1}\left(A_y / A_x\right),$$

$$A_x = A\cos(\theta) \quad A_y = A\sin(\theta),$$

where A is the magnitude of the vector. A vector magnitude is sometimes represented by absolute value notation: $A \equiv |A|$.

Notice that the inverse tangent gives a result which is ambiguous relative to adding or subtracting integer multiples of π. Thus the quadrant in which the angle lies must be resolved by independently examining the signs of A_x and A_y and choosing the appropriate value of θ.

To add two vectors, A and B, it is easiest to convert them to Cartesian component form. The components of the sum $C = A + B$ are then just the sums of the components:

$$C_x = A_x + B_x \quad C_y = A_y + B_y.$$

Subtraction of vectors is done similarly, e. g., if $A = C - B$, then

$$A_x = C_x - B_x \quad A_y = C_y - B_y.$$

A unit vector is a vector of unit length. One can always construct a unit vector from an ordinary vector by dividing the vector by its length: $n = A / |A|$. This division operation is carried out by dividing each of the vector components by the number in the denominator. Alternatively, if the vector is expressed in terms of length and direction, the magnitude of the vector is divided by the denominator and the direction is unchanged.

Unit vectors can be used to define a Cartesian coordinate system. Conventionally, i, j, and k indicate the x, y, and z axes of such a system. Note that i, j, and k are mutually perpendicular. Any vector can be represented in terms of unit vectors and its Cartesian components: $A = A_x i + A_y j + A_z k$. An alternate way to represent a vector is as a list of components: $A = (A_x, A_y, A_z)$. We tend to use the latter representation since it is somewhat more economical notation.

There are two ways to multiply two vectors, yielding respectively what are known as the dot product and the cross product. The cross product yields another vector while the dot product yields a number. Here we will discuss only the dot product.

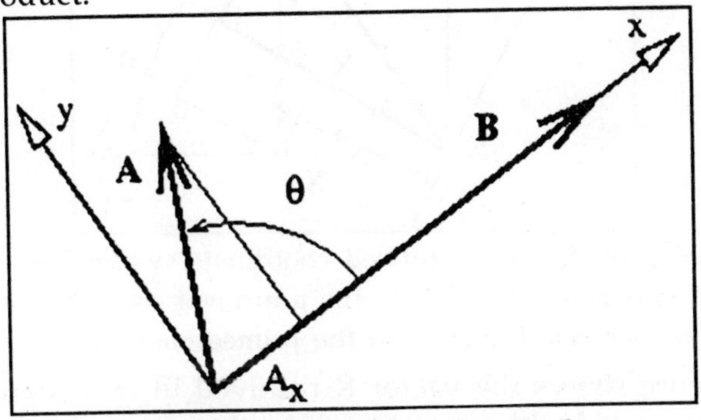

Fig. Definition Sketch for Dot Product

Given vectors A and B, and, the dot product of the two is defined as

$$A \cdot B \equiv |A||B| \cos\theta,$$

where θ is the angle between the two vectors. An alternate expression for the dot product exists in terms of the Cartesian components of the vectors:

$$A \cdot B = A_x B_x + A_y B_y.$$

It is easy to show that this is equivalent to the cosine form of the dot product when the axis lies along one of the vectors. Notice in particular that $A_x = |A|\cos\theta$, while $B_x = |B|$ and $B_y = 0$. Thus, $A \cdot B = |A|\cos\theta|B|$ in this case, which is identical to the form given in equation.

All that remains to be proven for equation to hold in general is to show that it yields the same answer regardless of how the Cartesian coordinate system is oriented relative to the vectors. To do this, we must show that $A_xB_x + A_yB_y = A_x'B_x' + A_y'B_y'$, where the primes indicate components in a coordinate system rotated from the original coordinate system.

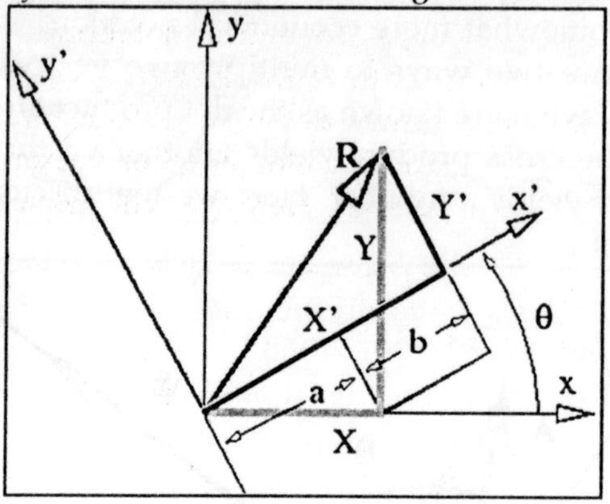

Fig. Definition figure for rotated coordinate system. The vector R has components X and Y in the unprimed coordinate system and components X′ and Y′ in the primed coordinate system.

Figure shows the vector R resolved in two coordinate systems rotated with respect to each other. From this figure it is clear that $X' = a + b$. Focusing on the shaded triangles, we see that $a = X\cos\theta$ and $b = Y\cos\theta$. Thus, we find $X' = X\cos\theta + Y\sin\theta$. Similar reasoning shows that. Substituting these and using the trigonometric identity results in

$$A_x'B_x' + A_y'B_y' = (A_x\cos\theta + A_y\sin\theta)(B_x\cos\theta + B_y\sin\theta)$$
$$+(-A_x\sin\theta + A_y\cos\theta)(-B_x\sin\theta + B_y\cos\theta)$$
$$= A_xB_x + A_yB_y$$

thus proving the complete equivalence of the two forms of the dot product as given by equations.

A numerical quantity which doesn't depend on which coordinate system is being used is called a scalar. The dot product of two vectors is a scalar. However, the components of a vector, taken individually, are not scalars, since the components change as the coordinate system changes. Since the laws of physics cannot depend on the choice of coordinate system being used, we insist that physical laws be expressed in terms of scalars and vectors, but not in terms of the components of vectors.

In three dimensions the cosine form of the dot product remains the same, while the component form is

$$\mathrm{A} \cdot \mathrm{B} = A_x B_x + A_y B_y + A_z B_z.$$

PLANE WAVES

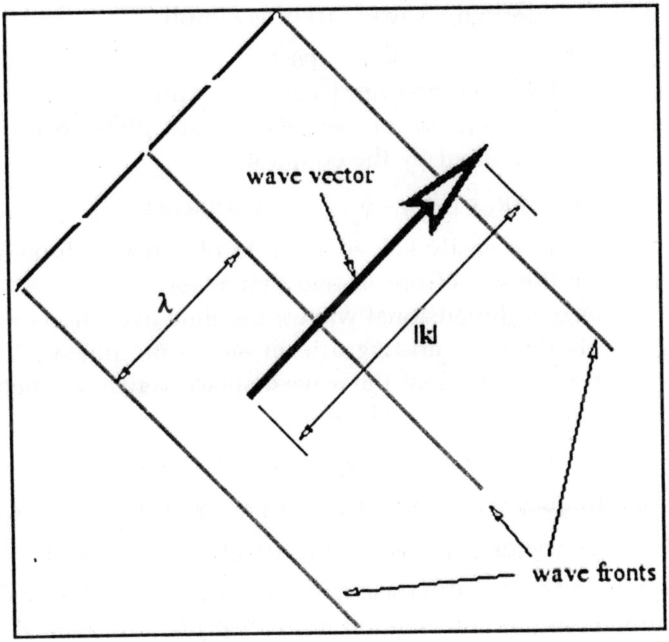

Fig. Definition sketch for a plane sine wave in two dimensions. The wave fronts are constant phase surfaces separated by one wavelength. The wave vector is normal to the wave fronts and its length is the wavenumber.

A plane wave in two or three dimensions is like a sine wave in one dimension except that crests and troughs aren't points, but form lines (2-D) or planes (3-D) perpendicular to the direction of wave propagation. Figure shows a plane sine wave in two dimensions. The large arrow is a vector called the wave vector, which defines :

- The direction of wave propagation by its orientation perpendicular to the wave fronts,
- The wavenumber by its length. We can think of a wave front as a line along the crest of the wave.

The equation for the displacement associated with a plane sine wave in three dimensions at some instant in time is

$$A(x,y,z) = \sin(\mathrm{k} \cdot \mathrm{x}) = \sin\left(k_x x + k_y y + k_z z\right).$$

Since wave fronts are lines or surfaces of constant phase, the equation defining a wave front is simply.

$$\mathrm{k} \cdot \mathrm{x} = const.$$

In the two dimensional case we simply set $k_z = 0$. Therefore, a wavefront, or line of constant phase ϕ in two dimensions is defined by the equation

$$\mathrm{k} \cdot \mathrm{x} = \mathrm{k}_\mathrm{x} x + k_y y = \phi \qquad \text{(two dimensions)}.$$

This can be easily solved for y to obtain the slope and intercept of the wavefront in two dimensions.

As for one dimensional waves, the time evolution of the wave is obtained by adding a term $-\omega t$ to the phase of the wave. In three dimensions the wave displacement as a function of both space and time is given by

$$A(x,y,z,t) = \sin\left(k_x x + k_y y + k_z z - \omega t\right).$$

The frequency depends in general on all three components of the wave vector. The form of this function, $\omega = \omega\left(k_x, k_y, k_z\right)$, which as in the one dimensional case is called the dispersion relation, contains information about the physical behaviour of the wave.

Some examples of dispersion relations for waves in two dimensions are as follows:

- Light waves in a vacuum in two dimensions obey

$$\omega = c\left(k_x^2 + k_y^2\right)^{1/2} \text{(light)},$$

- where is the speed of light in a vacuum.
- Deep water ocean waves in two dimensions obey

$$\omega = q^{1/2}\left(k_x^2 + k_y^2\right)^{1/4} \text{ (ocean waves)},$$

- where q is the strength of the Earth's gravitational field as before.
- Certain kinds of atmospheric waves confined to a vertical $x - z$ plane called gravity waves (not to be confused with the gravitational waves of general relativity) obey

$$\omega = \frac{Nk_x}{k_z} \qquad \text{(gravity waves)},$$

- where N is a constant with the dimensions of inverse time called the Brunt-Väisälä frequency.

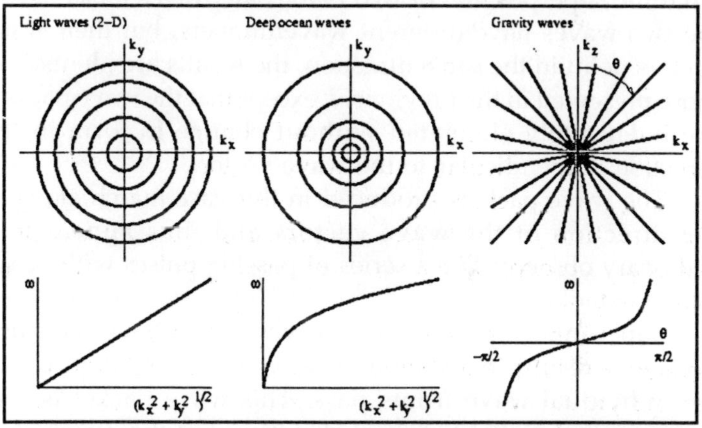

Fig. Contour Plots of the Dispersion Relations for Three Kinds of Waves in Two Dimensions

In the upper panels the curves show lines or contours along which the frequency takes w on constant values. Contours ar drawn for equally spaced values of w. For light and ocean waves the frequency depends only on the magnitude of the wave vector, whereas for gravity waves it depends only on the wave vector's direction, as defined by the angle q in the upper right panel. These dependences for each wave type are illustrated in the lower panels.

Contour plots of these dispersion relations are plotted in the upper panels of figure. These plots are to be interpreted like topographic maps, where the lines represent contours of constant elevation. In the case of figure, constant values of frequency are represented instead.

For simplicity, the actual values of frequency are not labeled on the contour plots, but are represented in the graphs in the lower panels. This is possible because frequency depends only on wave vector magnitude $\left(k_x^2 + k_y^2\right)^{1/2}$ for the first two examples, and only on wave vector direction for the third.

SUPERPOSITION OF PLANE WAVES

We now study wave packets in two dimensions by asking what the superposition of two plane sine waves looks like. If the two waves have different wavenumbers, but their wave vectors point in the same direction, the results are identical to those presented in the previously, except that the wave packets are indefinitely elongated without change in form in the direction perpendicular to the wave vector.

The wave packets produced in this case march along in the direction of the wave vectors and thus appear to a stationary observer like a series of passing pulses with broad lateral extent.

Superimposing two plane waves which have the same frequency results in a stationary wave packet through which the individual wave fronts pass. This wave packet is also elongated indefinitely in some direction, but the direction of elongation depends on the dispersion relation for the waves being considered.

One can think of such wave packets as steady beams, which guide the individual phase waves in some direction, but don't themselves change with time. By superimposing multiple plane waves, all with the same frequency, one can actually produce a single stationary beam, just as one can produce an isolated pulse by superimposing multiple waves with wave vectors pointing in the same direction.

Two Waves of Identical Wavelength

If the frequency of a wave depends on the magnitude of

the wave vector, but not on its direction, the wave's dispersion relation is called isotropic. In the isotropic case two waves have the same frequency only if the lengths of their wave vectors, and hence their wavelengths, are the same. The first two examples in figure satisfy this condition. In this section we investigate the beams produced by superimposing isotropic waves of the same frequency.

We superimpose two plane waves with wave vectors $k_1 = (-\Delta k_x, k_y)$ and $k_2 = (\Delta k_z, k_y)$. The lengths of the wave vectors in both cases are $k_1 = k_2 = \left(\Delta k_x^2 + k_y^2\right)^{1/2}$:

$$A = \sin\left(-\Delta k_x x + k_y y + -\omega t\right) + \sin\left(\Delta k_x x + k_y y - \omega t\right).$$

If $\Delta k_x \ll k_y$, then both waves are moving approximately in the y direction. An example of such waves would be two light waves with the same frequencies moving in slightly different directions.

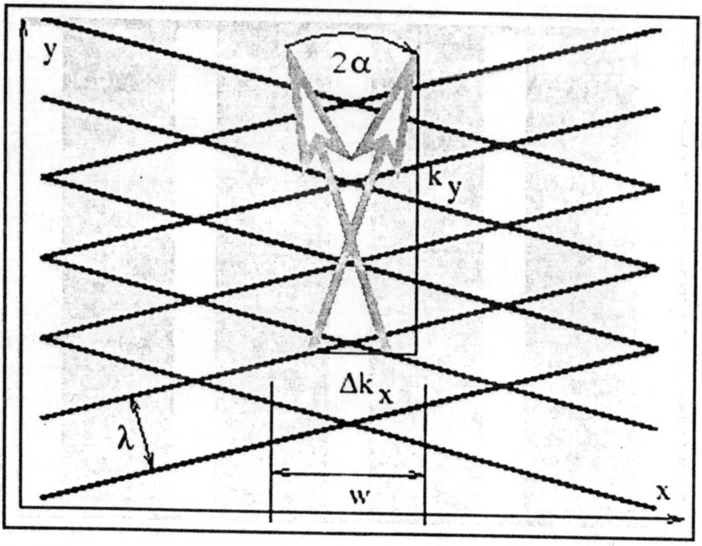

Fig. Wave Fronts and Wave Vectors of Two Plane Waves with the Same Wavelength but Oriented in Different Directions

The vertical bands show regions of constructive interference where wave fronts coincide. The vertical regions in between the bars have destructive interference, and hence define the lateral boundaries of the beams produced by the superposition. The components Dk_x and k_y of one of the wave vectors are shown.

Applying the trigonometric identity for the sine of the sum of two angles, equation) can be reduced to

$$A = 2\sin\left(k_y y - \omega t\right)\cos\left(\Delta k_x x\right).$$

This is in the form of a sine wave moving in the y direction with phase speed $c_{phase} = \omega / k_y$ and wavenumber k_y, modulated in the direction by a cosine function. The distance between regions of destructive interference in the direction tells us the width of the resulting beams, and is given by $\Delta k_x \omega = \pi$, so that

$$\omega = \pi / \Delta k_x.$$

Thus, the smaller Δk_x, the greater is the beam diameter.

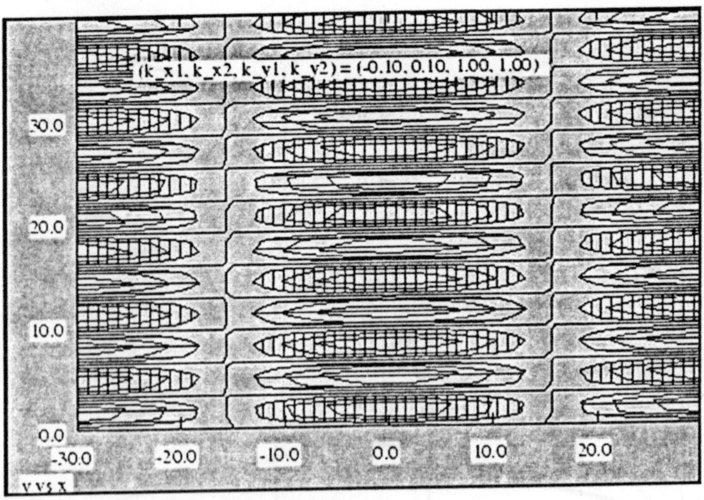

Fig. Example of Beams Produced by Two Plane Waves with the Same Wavelength Moving in Different Directions

The Wave Vectors of the Two Waves are $k = (\pm 0.1, 1.0)$. Regions of Positive Displacement are Illustrated by Vertical Hatching, while Negative Displacement has Horizontal Hatching.

Figure shows an example of the beams produced by superposition of two plane waves of equal wavelength oriented as in figure. It is easy to show that the transverse width of the resulting wave packet satisfies equation.

Two Waves of Differing Wavelength

In the third example of figure, the frequency of the wave depends only on the direction of the wave vector, independent of its magnitude, which is just the reverse of the case for an isotropic dispersion relation. In this case different plane waves with the same frequency have wave vectors which point in the same direction, but have different lengths.

More generally, one might have waves for which the frequency depends on both the direction and magnitude of the wave vector. In this case, two different plane waves with the same frequency would typically have wave vectors which differed both in direction and magnitude. We now investigate the superposition of non-isotropic waves with the same frequency.

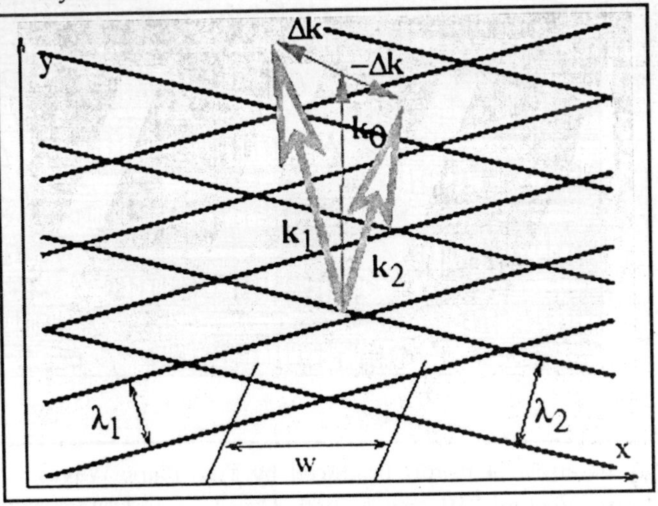

Fig. Wave Fronts and Wave Vectors (k_1 and k_2) of Two Plane Waves with Different Wavelengths Oriented in Different Directions

The slanted bands show regions of constructive

interference where wave fronts coincide. The slanted regions in between the bars have destructive interference, define the lateral limits of the beams produced by the superposition. The quantities k_0 and Dk are also shown.

Mathematically, we can represent the superposition of these two waves as a generalization of equation:

$$A = \sin\left[-\Delta k_x x + \left(k_y + \Delta k_y\right) y - \omega t\right]$$
$$+ \sin\left[\Delta k_x x + \left(k_y - \Delta k_y\right) y - \omega t\right].$$

In this equation we have given the first wave vector a y component $k_y + \Delta k_y$ while the second wave vector has $k_y - \Delta k_y$. As a result, the first wave has overall wavenumber $k_1 = \left[\Delta k_x^2 + \left(k_y + \Delta k_y\right)^2\right]^{1/2}$ while the second has $k_2 = \left[\Delta k_x^2 + \left(k_y - \Delta k_y\right)^2\right]^{1/2}$, so that $k_1 \neq k_2$.

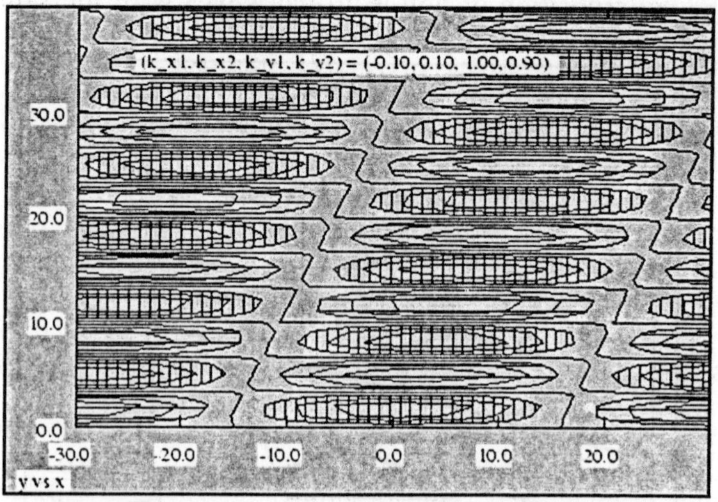

Fig. Example of Beams Produced by Two Plane Waves with Wave Vectors Differing in both Direction and Magnitude

The wave vectors of the two waves are $k_1 = (-0.1, 1.0)$ and $k_2 = (0.1, 0.9)$. Regions of positive displacement are illustrated by vertical hatching, while negative displacement has horizontal hatching.

Using the usual trigonometric identity, we write equation as

$$A = 2\sin\left(k_y y - \omega t\right)\cos\left(-\Delta k_x x + \Delta k_y y\right).$$

To see what this equation implies, notice that constructive interference between the two waves occurs when $-\Delta k_x x + \Delta k_y y = m\pi$, where m is an integer. Solving this equation for y yields $y = \left(\Delta k_x / \Delta k_y\right)x + m\pi / \Delta k_y$, which corresponds to lines with slope $\Delta k_x / \Delta k_y$.

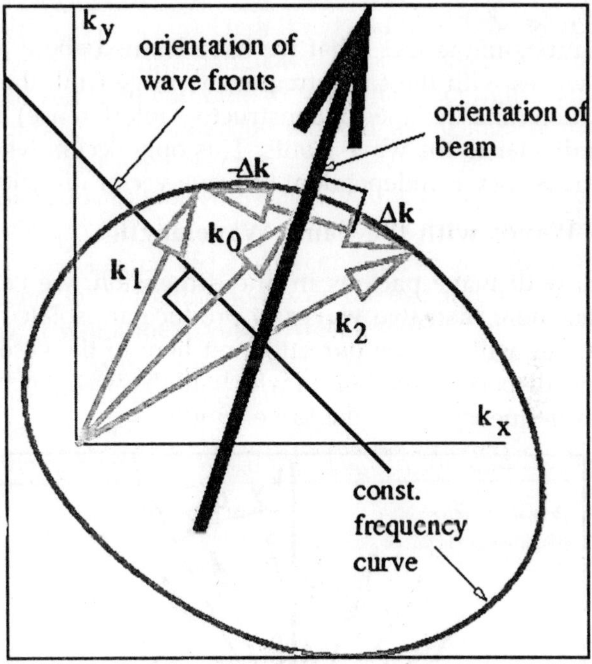

Fig. Illustration of Factors Entering the Addition of Two Plane Waves with the Same Frequency

These lines turn out to be perpendicular to the vector difference between the two wave vectors, $k_2 - k_1 = \left(2\Delta k_x, -2\Delta k_y\right)$. The easiest way to show this is to note that this difference vector is oriented so that it has a slope $-\Delta k_y/\Delta k_x$. Comparison with the $\Delta k_x/\Delta k_y$ slope of the lines of constructive interference indicates that this is so.

An example of the production of beams by the superposition of two waves with different directions and wavelengths is shown in figure. Notice that the wavefronts are still horizontal, but that the beams are not vertical, but slant to the right.

The wave fronts are perpendicular to the vector average of the two wave vectors, $k_2 = (k_1 + k_2)/2$, while the lines of constructive interference, which define the beam orientation, are oriented perpendicular to the difference between these two vectors, $k_2 - k_1$.

Figure summarizes what we have learned about adding plane waves with the same frequency. In general, the beam orientation (and the lines of constructive interference) are not perpendicular to the wave fronts. This only occurs when the wave frequency is independent of wave vector direction.

Many Waves with the Same Wavelength

As with wave packets in one dimension, we can add together more than two waves to produce an isolated wave packet. We will confine our attention here to the case of an isotropic dispersion relation in which all the wave vectors for a given frequency are of the same length.

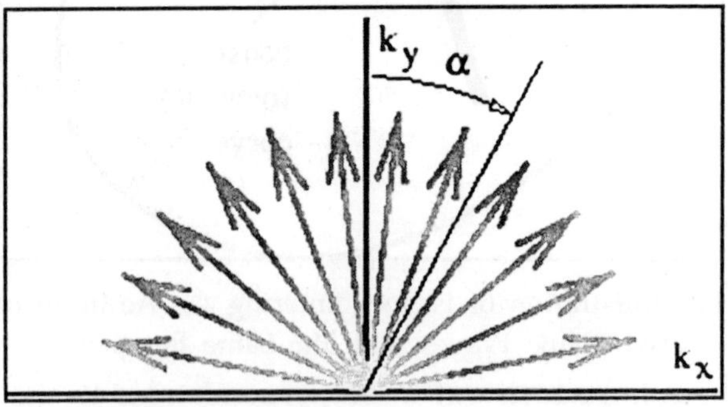

Fig. Illustration of Wave Vectors of Plane Waves which Might be Added Together

Figure shows an example of this in which wave vectors

of the same wavelength but different directions are added together. Defining α_i as the angle of the th wave vector clockwise from the vertical, we could write the superposition of these waves at time $t = 0$ as

$$A = \sum_i A_i \sin\left(k_{xi}x + k_{yi}y\right)$$
$$= \sum_i A_i \sin\left[kx\sin(\alpha_i) + ky\cos(\alpha_i)\right]$$

where we have assumed that $k_{xi} = k\sin(\alpha_i)$ and $k_{yi} = k\cos(\alpha_i)$. The parameter is the magnitude of the wave vector and is the same for all the waves. Let us also assume in this example that the amplitude of each wave component decreases with increasing $|\alpha_i|$:

$$A_i = \exp\left[-(\alpha_i/\alpha_{\max})^2\right].$$

The exponential function decreases rapidly as its argument becomes more negative, and for practical purposes, only wave vectors with $|\alpha_i| \le \alpha_{\max}$ contribute significantly to the sum. We call $\alpha_{\max}$ the spreading angle.

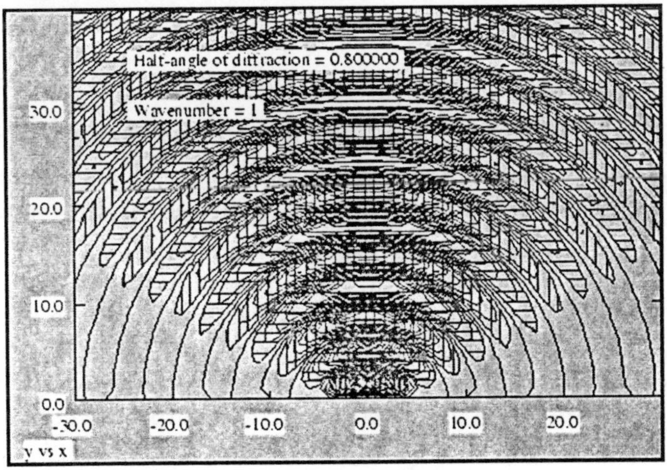

Fig. Plot of the Displacement Field A(x, y) from Equation for $\alpha_{\max}$ and $k = 1$.

Figure shows what $A(x,y)$ looks like when $\alpha_{max} = 0.8$ radians and $k=1$. Notice that for $y=0$ the wave amplitude is only large for a small region in the range $-4 < x < 4$. However, for $y > 0$ the wave spreads into a broad semicircular pattern.

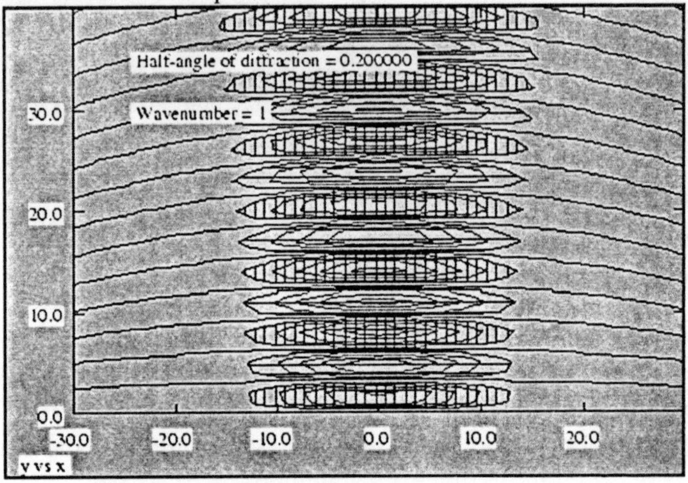

Fig. Plot of the Displacement Field $A(x,y)$ from Equation for $\alpha_{max} = 0.2$ and $k=1$.

Figure shows the computed pattern of $A(x,y)$ when the spreading angle $\alpha_{max} = 0.2$ radians. The wave amplitude is large for a much broader range of x at $y = 0$ in this case, roughly $-12 < x < 12.$. On the other hand, the subsequent spread of the wave is much smaller than in the case of figure.

We conclude that a superposition of plane waves with wave vectors spread narrowly about a central wave vector which points in the y direction produces a beam which is initially broad in but for which the breadth increases only slightly with increasing y. However, a superposition of plane waves with wave vectors spread more broadly produces a beam which is initially narrow x in but which rapidly increases in width as y increases.

The relationship between the spreading angle α_{max} and the initial breadth of the beam is made more understandable by comparison with the results for the two-wave superposition discussed at the beginning of this section. As indicated by equation, large values of k_x, and hence α, are associated with small wave packet dimensions in the direction and vice versa.

The superposition of two waves doesn't capture the subsequent spread of the beam which occurs when many waves are superimposed, but it does lead to a rough quantitative relationship between α_{max} (which is just $\tan^{-1}(k_x/k_y)$ in the two wave case) and the initial breadth of the beam. If we invoke the small angle approximation for $\alpha = \alpha_{max}$ so that , then and equation can be written $w = \pi / k_x \approx \pi/(k\alpha_{max}) = \lambda/(2\alpha_{max})$.

Thus, we can find the approximate spreading angle from the wavelength of the wave and the initial breadth of the beam w:

$$\alpha_{max} = \lambda/(2w) \text{ (signal slit spreading angle)}$$

Diffraction Through a Single Slit

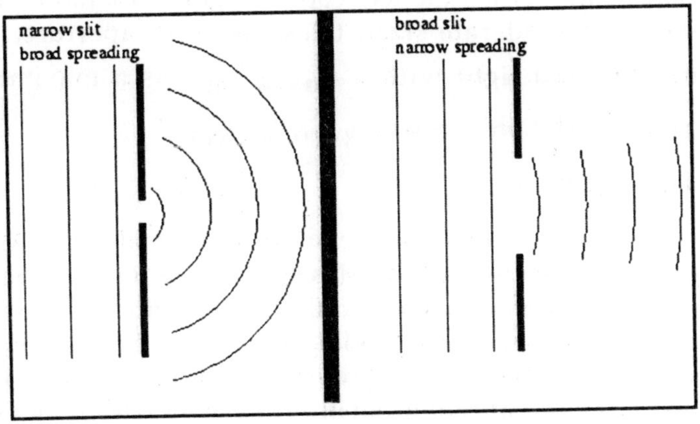

Fig. Schematic Behaviour when a Plane Wave Impinges on a Narrow Slit and a Broad Slit

How does all of this apply to the passage of waves through a slit? Imagine a plane wave of wavelength λ impingent on a barrier with a slit. The barrier transforms the plane wave with infinite extent in the lateral direction into a beam with initial transverse dimensions equal to the width of the slit.

The subsequent development of the beam is illustrated in figures, and schematically in figure. In particular, if the slit width is comparable to the wavelength, the beam spreads broadly as in figure. If the slit width is large compared to the wavelength, the beam doesn't spread as much. Equation gives us an approximate quantitative result for the spreading angle if is interpreted as the width of the slit.

One use of the above equation is in determining the maximum angular resolution of optical instruments such as telescopes. The primary lens or mirror can be thought of as a rather large "slit". Light from a distant point source is essentially in the form of a plane wave when it arrives at the telescope.

However, the light passed by the telescope is no longer a plane wave, but is a beam with a tendency to spread. The spreading angle α_{max} is given by equation, and the telescope cannot resolve objects with an angular separation less than α_{max}. Replacing with the diameter of the lens or mirror in equation thus yields the telescope's angular resolution. For instance, a moderate sized telescope with aperture l m observing red light with $\lambda \approx 6 \times 10^{-7}$ m has a maximum angular resolution of about 3×10^{-7} radians.

TWO SLITS

Let us now imagine a plane sine wave normally impingent on a screen with two narrow slits spaced by a distance. Since the slits are narrow relative to the wavelength of the wave impingent on them, the spreading angle of the beams is large and the diffraction pattern from each slit individually is a cylindrical wave spreading out in all directions. The cylindrical waves from the two slits interfere, resulting in oscillations in wave intensity at the screen on the right side of figure.

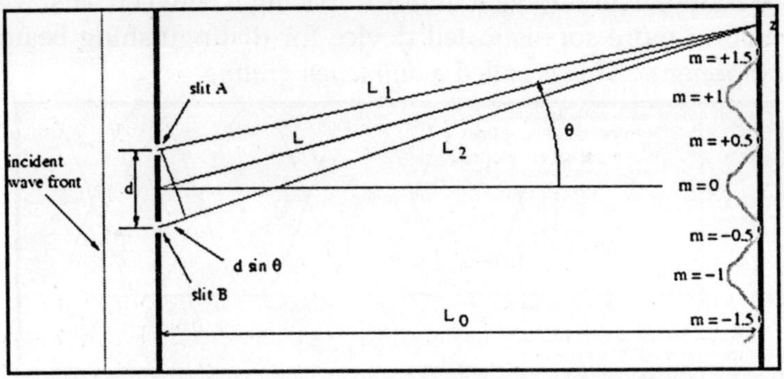

Fig. Definition Sketch for the Double Slit. Light passing through Slit B Travels an Extra Distance to the Screen equal to $d\sin\theta$ compared to Light Passing Through Slit A.

Constructive interference occurs when the difference in the paths traveled by the two waves from their originating slits to the screen, $L_2 - L_1$, is an integer multiple of the wavelength $\lambda : L_2 - L_1 = m\lambda$.

If $L_0 \gg d$, the lines L_1 and L_2 are nearly parallel, which means that the narrow end of the dark triangle in figure has an opening angle of θ. Thus, the path difference between the beams from the two slits is $L_2 - L_1 = d\sin\theta$. Substitution of this into the above equation shows that constructive interference occurs when

$$d\sin\theta = m\lambda, m = 0, \pm 1, \pm 2, \ldots \text{(two slit interference)}.$$

Destructive interference occurs when m is an integer plus 1/2. The integer is called the interference order and is the number of wavelengths by which the two paths differ.

DIFFRACTION GRATINGS

Since the angular spacing Δθ of interference peaks in the two slit case depends on the wavelength of the incident wave, the two slit system can be used as a crude device to distinguish between the wavelengths of different components of a non-sinusoidal wave impingent on the slits. However, if more slits are added, maintaining a uniform spacing *d* between slits, we obtain a more sophisticated device for distinguishing beam components. This is called a *diffraction grating*.

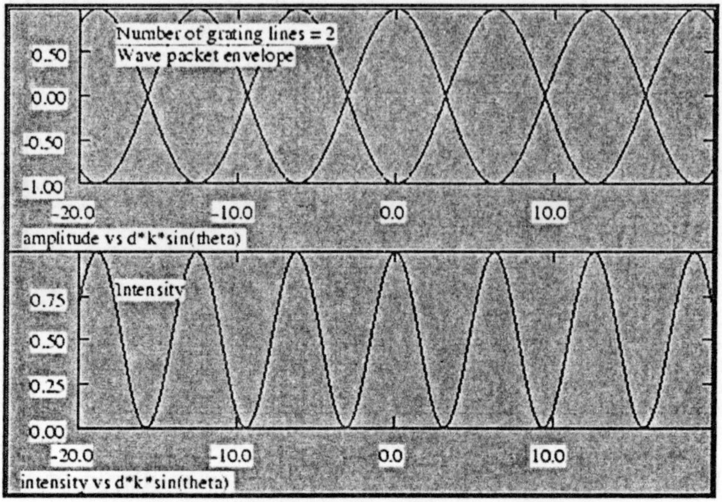

Fig. Amplitude And Intensity of Interference Pattern from a Diffraction Grating with 2 Slits.

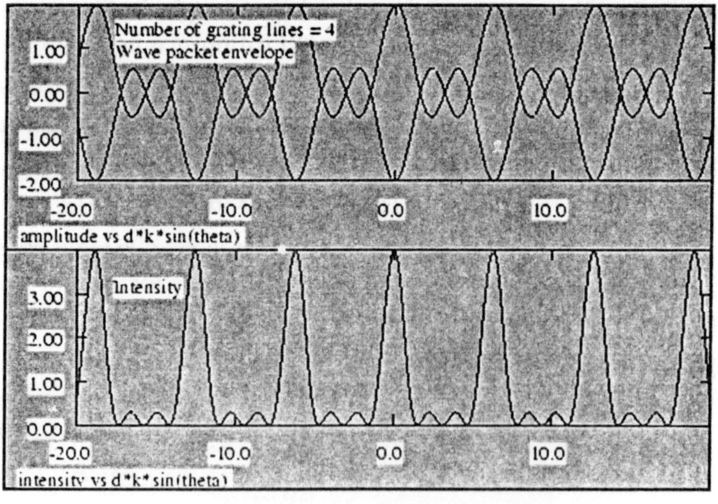

Fig. Amplitude and Intensity of Interference Pattern from a Diffraction Grating with 4 Slits.

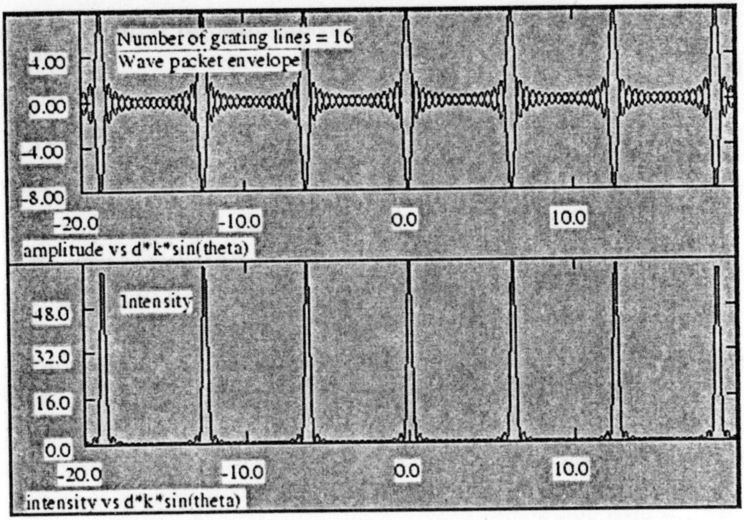

Fig. Amplitude and Intensity of Interference Pattern from a Diffraction Grating with 16 Slits.

Figures show the amplitude and intensity of the diffraction pattern for gratings with 2, 4, and 16 slits respectively. Notice how the interference peaks remain in the same place but increase in sharpness as the number of slits increases.

The width of the peaks is actually related to the overall width of the grating, $w = nd$, where is the number of slits. Thinking of this width as the dimension of large single slit, the single slit equation, $\alpha_{max} = \lambda/(2w)$, tells us the angular width of the peaks.

Whereas the angular width of the interference peaks is governed by the single slit equation, their angular positions are governed by the two slit equation. Let us assume for simplicity that $|\theta| \ll 1$ so that we can make the small angle approximation to the two slit equation, $m\lambda = d\sin\theta \approx d\theta$, and ask the following question: How different do two wavelengths have to be in order that the interference peaks from the two waves not overlap? In order for the peaks to be distinguishable, they should be separated in θ by an angle

$\Delta\theta = m\Delta\lambda / d$, which is greater than the angular width of each peak, α_{max}:

$$\Delta\theta > \alpha_{max}.$$

Substituting in the above expressions for $\Delta\theta$ and α_{max} and solving for $\Delta\theta$, we get $\Delta\lambda > \lambda / (2mn)$, where $n = w/d$ is the number of slits in the diffraction grating. Thus, the fractional difference between wavelengths which can be distinguished by a diffraction grating depends solely on the interference order and the number of slits *n* in the grating:

$$\frac{\Delta\lambda}{\lambda} > \frac{1}{2mn}.$$

Chapter 9

Fresnel's Laws in Discrete Optical Media

When light crosses the interface between two dissimilar media, it experiences refraction as well as partial reflection. In dielectric media, these phenomena are described by the famous laws of Snell and Fresnel. When deriving these mathematical expressions, it is implicitly assumed that the media on both sides of the interface are continuous. Hence, assuming a fixed wavelength, they are characterized by a single quantity: the refractive index.

However, in modern physics, a variety of more sophisticated systems such as periodically structured media, e.g. photonic crystals, constitute an important aspect in many research fields. During recent years, a subclass of such periodic media has gained much interest: arrays of evanescently coupled waveguides, which are a representation of so-called discrete systems.

During recent years, surfaces of and interfaces between waveguide arrays have attracted rapidly growing interest, due to their strong impact on the evolution of light. Light propagation in finite waveguide lattices was modeled in one- (1D) and two-dimensional (2D) geometries. The tunneling escape from a boundary waveguide was investigated in close analogy to quantum mechanics. Furthermore, the conditions for existence of localized linear modes at defects and interfaces were analyzed in detail.

The research on discrete propagation in waveguide arrays gained further stimulus, when the existence of localized nonlinear modes at surfaces and interfaces in waveguide arrays was predicted. These phenomena were experimentally demonstrated in 1D and only recently in 2D.

It turned out that light propagation at the interface of two waveguide lattices exhibits a variety of peculiarities as a direct consequence of the discrete nature of the medium. However, to date the transition of light through an interface between two dissimilar waveguide arrays was not analyzed. This is somewhat surprising since one can expect that interfaces also have a strong impact on the refraction, reflection and transmission properties of light, culminating in modifications of the laws of Snell and Fresnel with respect to the continuous case.

We bridge this gap and present the discrete version of Snell's law of refraction as well as reflection and transmission coefficients for plane waves passing an interface between two dissimilar waveguide arrays. It will be shown that reflection and refraction are determined by the change of the coupling and the propagation constants at the interface.

THE DISCRETE SNELL'S LAW

In the so-called coupled mode approximation the transverse profile of the propagating modes inside the individual guides is assumed to remain constant, so that only the amplitudes evolve along z. Then one can derive a set of coupled ordinary homogeneous first-order differential equations

$$i\frac{d}{dz}\varphi_m + c(\varphi_{m-1} + \varphi_{m+1}) = 0.$$

Here, φ_m is the amplitude in the mth waveguide and c is the coupling constant, denoting the coupling strength between the individual lattice sites. For reasons of simplicity, the propagation constant β_0 has been removed by a phase transformation. The normalized eigenfunctions of this periodic system are the so-called discrete Bloch waves:

$$\varphi_m = e^{i\beta z}e^{ikm}.$$

The quantity β is the longitudinal propagation constant and κ denotes the transverse propagation constant multiplied by the waveguide spacing Δ corresponding to a phase difference between adjacent waveguides. Inserting equation, one finds β entirely defined by the dispersion relation

$$\beta = 2c\cos\kappa$$

describing a band structure due to the system's periodicity. All Bloch waves have to fulfill equation, otherwise they cannot propagate and decay exponentially.

However, equation describes an infinite system with identical and equally spaced waveguides. Hence, in the presence of an interface, one has to modify these equations. Assuming an interface as depicted in figure, the coupled mode equations take the form

$$(m<0): i\frac{d}{dz}\varphi_m + c_1(\varphi_{m-1} + \varphi_{m+1}) = 0,$$

$$(m=0): \left(i\frac{d}{dz} + \Theta\right)\varphi_0 + c_1\varphi_{-1} + c_2\varphi_1 = 0,$$

$$(m>0): \left(i\frac{d}{dz} + \delta\beta\right)\varphi_m + c_2(\varphi_{m-1} + \varphi_{m+1}) = 0,$$

where $m = 0$ is the position of the interface. The quantities c_1 and c_2 are the coupling constants on the two sides of the interface, whereas δβ is the difference of the propagation constants in the isolated waveguides between these two parts of the array. If the guide $m = 0$ exhibits a detuning δβ, the relation Θ = δβ holds, otherwise Θ = 0.

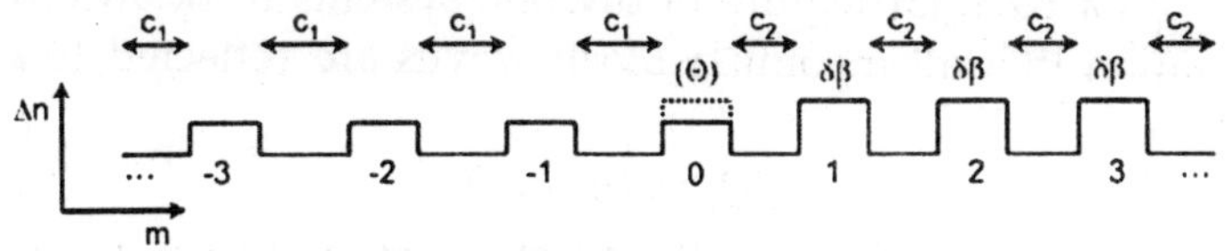

Fig. Schematic Representation of an Interface in a Waveguide Array, which is Induced by a change of the Coupling Constant and the Propagation Constants of the Guides.

For Θ = 0, one obtains two different dispersion relations for both regions with $n < 0$ and $n > 0$, respectively, which read

$$(m<0): \beta_1 = 2c_1\cos\kappa_1,$$

$$(m<0): \beta_2 = \delta\beta + 2c_2\cos\kappa_2.$$

Due to translational invariance in the *z*-direction, the propagation constant β must be conserved in the whole array. Hence, the relation $\beta_1 = \beta_2 = \beta$ holds, yielding the expression

$$\frac{\cos\kappa_2}{\cos\kappa_1} = \frac{c_1}{c_2} - \frac{\delta\beta}{2c_2\cos\kappa_1}.$$

This is the discrete equivalent of the well-known Snell's law, which describes the refraction of a propagating plane wave at an interface between two discrete media.

Following this analogy, in the discrete case in equation one has two parameters describing the refraction: the ratio of the coupling constants c_1/c_2 as well as the ratio $\delta\beta/c_2$, which is determined by an additional difference of the waveguide profiles.

This is in strong contrast to continuous media, where only the ratio of the refractive indices of both materials affects the refraction at the interface. The actual angle $\alpha_{1,2}$ of the propagation direction of the individual Bloch waves in the respective array region can be easily calculated by

$$\tan\alpha_{1.2} = \Delta_{1,2}\frac{\partial\beta}{\partial\kappa_{1,2}} = -2\Delta_{1,2}c_{1,2}\sin\kappa_{1,2}.$$

Furthermore, it is evident that only such waves can pass the interface for which $|\cos\kappa_2|d \le 1$ holds. Otherwise the Bloch waves experience total internal reflection, since they cannot penetrate the region behind the interface and decay exponentially in this region in the form of evanescent waves.

As a peculiar feature of discrete systems it follows from equation that all incoming Bloch waves are reflected totally when

$$|\delta\beta| \ge 2(c_1 + c_2).$$

In this particular case, the interface acts as a perfect mirror. Since the coupling constants and the detuning are a function of the wavelength, condition will only be satisfied for a certain range of wavelengths.

Hence, this feature may find an application as a spectral filter, where only certain wavelengths can penetrate the interface.

The occurrence of total reflection can be illustrated by means of the band structure. We consider the bands on both sides of the interface according to equations. Since only waves with a positive value of the Bloch vector reach the interface, we examine the bands in the interval [0; π].

Two examples are shown in figure. In figure the band is shifted up behind the interface because a positive value for δβ is assumed. In this case, no propagation constant exists inside the right part of the array for Bloch waves from the bottom of the band of the left part.

Therefore, such a wave would be completely reflected. An analogous situation occurs for a shrinking of the band due to a decrease of the coupling constant, where total reflection appears if the incoming wave travels at the top or bottom of the respective band.

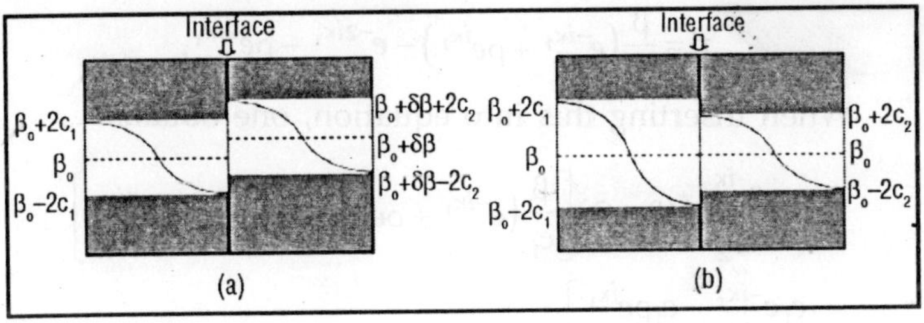

Fig. Bandstructure on Both Sides of the Interface. In (a) the Band of Both Sides Exhibits the Same width ($c_1 = c_2$) but are Shifted ($\delta\beta > 0$). In (b) the bands are different in width ($c_1 > c_2$), but are not shifted ($\delta\beta = 0$).

REFLECTION AND TRANSMISSION COEFFICIENTS

When the Bloch waves reach the interface, they split into a reflected and transmitted part. The resulting amplitudes of the Bloch waves can be characterized by reflection and transmission coefficients, in close analogy to the Fresnel's formulae at interfaces between continuous media.

To calculate these coefficients, for the propagating Bloch waves one uses

$$(m<0):\varphi_m=\left(e^{i\kappa_1 m}+\rho e^{-i\kappa_1 m}\right)e^{i\beta z},$$

$$(m=0):\varphi_m=\tau' e^{i\beta z},$$

$$(m>0):\varphi_m=\tau e^{i\kappa_2 m}e^{i\beta z}.$$

The coefficient ρ denotes the reflection strength, whereas τ' and τ are the transmission coefficients. Since the guide at $m = 0$ can be assumed as a singular defect, one has to consider the transmission into this defect and the $m > 0$-region separately. Inserting equations, one obtains

$$\alpha_{-1}:-\beta\left(e^{-i\kappa_1}+\rho e^{i\kappa_1}\right)+c_1 e^{-2i\kappa_1}+c_1\rho e^{2i\kappa_1}+c_1\tau'=0,$$

$$\alpha_0:(-\beta+\Theta)\tau'+c_1 e^{-i\kappa_1}+c_1\rho e^{i\kappa_1}+c_2\tau e^{i\kappa_2}=0,$$

$$\alpha_1:(-\beta+\delta\beta)\tau e^{i\kappa_2}+c_2\tau'+c_2\tau e^{2i\kappa_2}=0.$$

Equation directly yields

$$\tau'=\frac{\beta}{c_1}\left(e^{-i\kappa_1}+\rho e^{i\kappa_1}\right)-e^{-2i\kappa_1}-\rho e^{2i\kappa_1}.$$

When inserting this into equation, one obtains

$$\tau'=\frac{e^{-i\kappa_2}}{c_2}\left[(\beta-\Theta)\left[\frac{\beta}{c_1}\left(e^{-i\kappa_1}+\rho e^{i\kappa_1}\right)-e^{-2i\kappa_1}-\rho e^{2i\kappa_1}\right]\right.$$

$$\left.-c_1 e^{-i\kappa_1}-c_1\rho e^{i\kappa_1}\right].$$

Using equations, one can resolve equation, resulting in

$$\rho=\frac{\left(\left[-\beta+\delta\beta+c_2 e^{i\kappa_2}\right]/c_2\right)\left((\beta-\Theta)\left[e^{-i\kappa_1}-(\beta/c_1)\right]+c_1\right)}{\left(\left[-\beta+\delta\beta+c_2 e^{i\kappa_2}\right]/c_2\right)\left((\beta-\Theta)\left[(\beta/c_1)-e^{-i\kappa_1}\right]-c_1\right)}$$

$$\frac{-(c_2/c_2)\beta+c_2 e^{-i\kappa_1}}{+(c_2/c_1)\beta-c_2 e^{i\kappa_1}}e^{-2i\kappa_1},$$

where β has to fulfill the dispersion relations. Equations denote the transmission and reflection coefficients for a Bloch wave crossing an interface as depicted in figure from left to the right. Three examples for the coefficients in dependence of the Bloch vector of the incoming wave are depicted in figure.

An interesting issue is the symmetry of the reflection and transmission coefficients around $\kappa_1=\frac{\pi}{2}$, since due to equation

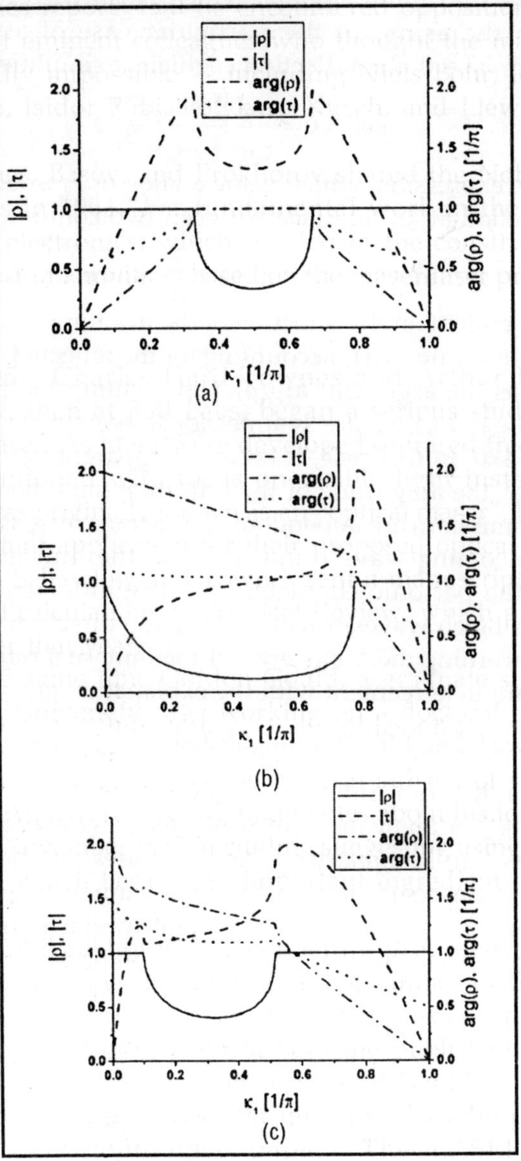

Fig. Absolute Value (solid) and Argument (dashed) of the Reflection (red) and Transmission (blue) Coefficients for (a) db = 0 cm^{-1}, $c_1 = 1$ cm^{-1} and $c_2 = 0.5$ cm^{-1}, (b) db = 0.5 cm^{-1}, $c_1 = c_2 = 1$ cm^{-1} and (c) db = 0.9 cm^{-1}, $c_1 = 1$ cm^{-1} and $c_2 = 0.5$ cm^{-1}.

this is the maximum angle at which Bloch waves propagate through the array. In the particular case of a vanishing detuning $\delta\beta = 0$, the reflection coefficient simplifies to

$$\rho|\delta\beta = 0 = \frac{c_1 e^{-i\kappa_1} - c_2 e^{i\kappa_2}}{c_2 e^{ik_2} - c_1 e^{-ik_1}}.$$

This expression, which is the analog of Fresnel's formula for reflection at an interface between two continuous media, is always symmetric around $\kappa_1 = \frac{\pi}{2}$, since the reflection is only dependent on the direction of the incoming wave. Hence, the absolute value $|\rho|$, accounting for the reflected amplitude, as well as the argument $\arg(\rho)$, representing the phase shift due to the reflection, are symmetric. This symmetry is broken when $\delta\beta \neq 0$. In this case, it becomes apparent that the phase as well as the amplitude of the reflection (and, hence, also of the transmission) are dependent on the transverse wave vector of the incoming wave rather than on the direction. This is due to the periodic character of waveguide arrays, where different Bloch waves travel in the same direction.

Concerning the occurrence of total internal reflection, the results are in agreement with the explanation given to figure. It might be astonishing that the transmitted wave does not vanish in this case. However, this becomes clear keeping in mind that the transmitted wave has to be either a Bloch wave (for real κ_2) or an evanescent wave decaying exponentially in the case of an imaginary κ_2. As both cases are included, the transmission coefficient τ does not vanish even for total reflection $|\pi| = 1$. To derive a quantity that represents the power conservation, it is useful to examine the energy flow inside the array. For an infinite array, it follows from equation

$$i\overset{*}{\varphi}_m \frac{d}{dz}\varphi_m + c(\varphi_{m-1} + \varphi_{m+1})\overset{*}{\varphi}_m = 0,$$

$$-i\varphi_m \frac{d}{dz}\overset{*}{\varphi}_m + c(\overset{*}{\varphi}_{m-1} + \overset{*}{\varphi}_{m+1})\varphi_m = 0.$$

Subtracting the second from the first equation yields

$$\frac{d}{dz}|\varphi_m|^2 + ic\left[(\varphi_{m-1} + \varphi_{m+1})\overset{*}{\varphi}_m - (\overset{*}{\varphi}_{m-1} + \overset{*}{\varphi}_{m+1})\varphi_m\right],$$

which denotes the power exchange between a respective guide and its neighbors. The power inside a cluster of guides extending from $-M$ to M is given by

$$Q = \sum_{-M}^{M} |\varphi_m|^2 .$$

Power can escape the cluster only via the two outermost waveguides. This is reflected in the corresponding equation

$$\frac{d}{dz} Q = ic\left[\left(\varphi^*_{-M}\varphi_{-M-1} - \varphi_{-M}\varphi^*_{-M} + \varphi^*_{M}\varphi_{M+1} - \varphi_{M}\varphi^*_{M+1}\right],$$

where only the amplitudes of the outermost guides and their neighbors outside the cluster contribute. To determine the power flow one assumes a cluster including all guides from $m = -\infty$ to $m = 0$. The power change of this area, which corresponds to the (md ≤ 0)-region of the array, is given by

$$\frac{d}{dz} Q\Big|_{m \leq o} = ic_2\left[\varphi^*_o\varphi_1 - \varphi_o\varphi^*_1\right] = 2c_2\Im\left\{\varphi_o\varphi^*_1\right\}.$$

Assuming plane waves $\varphi_m = e^{i\hat{e}_2 m}$, this equation becomes

$$\frac{d}{dz} Q\Big|_{m \leq o} = \begin{cases} -2c_2 \sin\kappa_2 (\text{ if } \kappa_2 \text{ is real}), \\ o \quad (\text{if } \kappa_2 \text{ is imaginary}). \end{cases}$$

As expected, no power flow exists for evanescent waves (κ_2 imaginary). Assuming that the power transported by the reflected and transmitted wave equals the power of the incoming wave, the coefficients have to fulfill

$$c_1 \sin\kappa_1 = |\rho|^2 c_1 \sin\kappa_1 + |\tau|^2 c_2 \sin\kappa_2 .$$

Using equation, one finally arrives at

$$1 = |\rho|^2 + \frac{1}{\sin\kappa_1}\sqrt{\left(\frac{c_2}{c_1}\right)^2 - \left(\cos\kappa_1 - \frac{\delta\beta}{2c_1}\right)^2}\, |\tau|^2 ,$$

which represents the conservation of power inside the array. Hence, the quantities

$$R = |\rho|^2 ,$$

$$T = \frac{1}{\sin\kappa_1}\sqrt{\left(\frac{c_2}{c_1}\right)^2 - \left(\cos\kappa_1 - \frac{\delta\beta}{2c_1}\right)^2}\, |\tau|^2 ,$$

represent the reflection and transmission coefficients for the power. As expected, the transmission T grows with

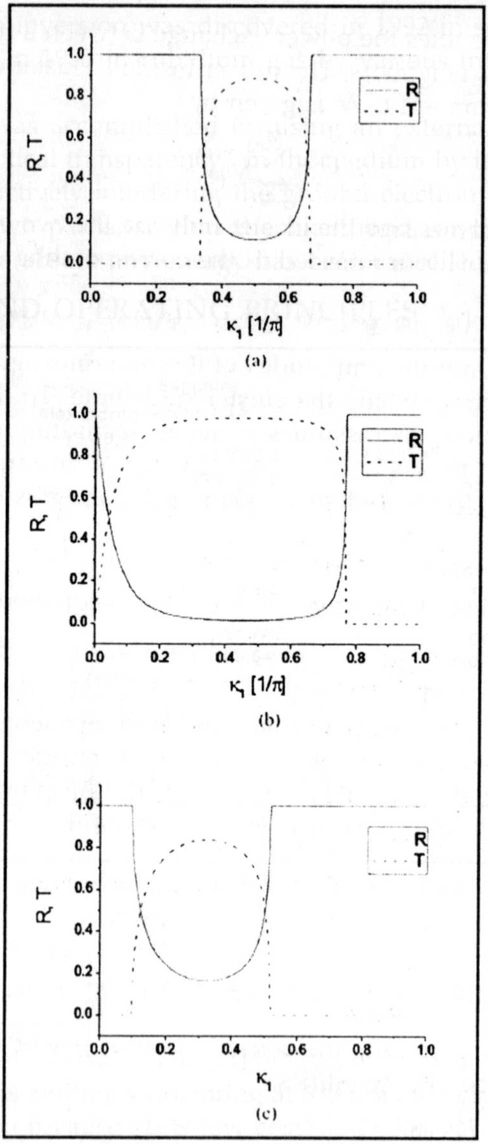

Fig. Energy Reflection (red) and Transmission (blue) Coefficients for (a) $\delta\beta = 0\ cm^{-1}$, $c_1 = 1\ cm^{-1}$ and $c_2 = 0.5\ cm^{-1}$, (b) $\delta\beta = 0.5\ cm^{-1}$, $c_1 = c_2 = 1\ cm^{-1}$ and (c) $\delta\beta = 0.9\ cm^{-1}$, $c_1 = 1\ cm^{-1}$ and $c_2 = 0.5\ cm^{-1}$.

decreasing reflection R and vice versa, whereas their sum is constant. The corresponding values for R and T from figure are depicted in figure.

ANALYTICAL RESULTS

The coupled mode equations were solved numerically. The first point under investigation will be the discrete Snell's law. Since this expression is valid for every single Bloch wave, it is useful to analyse this for a broad excitation covering at least five waveguides. In this case, the Fourier spectrum is narrow, so that only a small number of Bloch waves will be excited. Investigating the evolution of the light inside the array around $\kappa = \pi/2$ diffraction is small due to the significantly reduced curvature of the band in this area . Therefore almost no broadening of the beam appears.

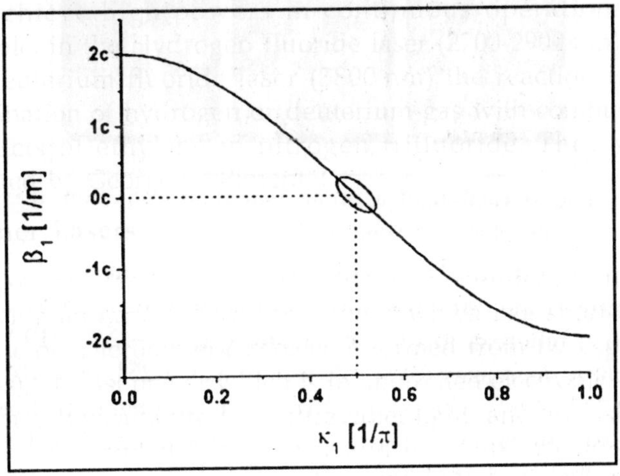

Fig. In the Vicinity of the Normalized Transverse Wave Vector $\kappa = \frac{\pi}{2}$ the Second Derivative of the Longitudinal Wave Vector Vanishes, which is Equivalent to Diffraction-free Propagation.

In the array under consideration, on the left side ($m < 0$) the coupling was set to $c_1 = 1\ \text{cm}^{-1}$ for a spacing of $_{\Delta 1} = 15\ \mu\text{m}$ and on the right side ($m > 0$) to $c_2 = 2\ \text{cm}^{-1}$, according to

$_{\Delta 2}$ = 11 μm . For these values, in region 1 one obtains a propagation angle of $\alpha_1 = 0.18°$, whereas in the other region $\alpha_2 = 0.25°$, which is shown in figure.

The picture changes for an additional detuning of the refractive index in the $m > 0$ region. For a detuning of $\delta n = -1.5\times10^{-4}$, the angle of propagation inside this region reduces to $\alpha_2 = 0.23°$, whereas for an index detuning of $\delta n = -2.7\times10^{-4}$, the angle of propagation reduces to $\alpha_2 = 0.18°$, so that no refraction occurs (but still a reflection), which is shown in figure.

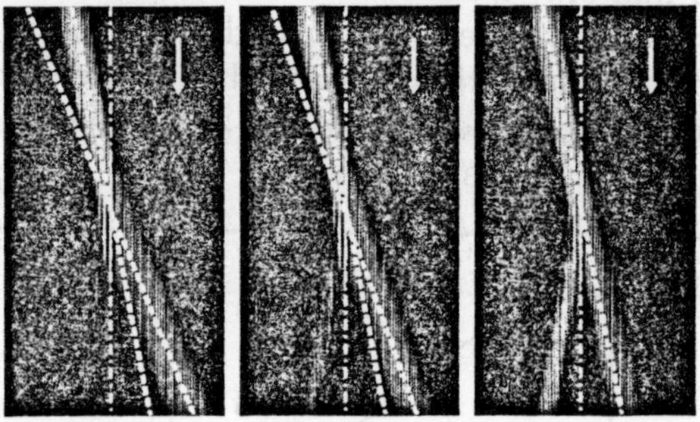

(a) (b) (c)

Fig. Refraction of a broad Gaussian excitation at an interface between $c_1 = 1\ \text{cm}^{-1}$ and $c_2 = 2\ \text{cm}^{-1}$.

In (a) no additional detuning is assumed, in (b) the detuning of the refractive index in the ($m > 0$) region is $dn = -1.5\times10^{-4}$, whereas in (c) the detuning is $dn = 2.7 \times 10^{-4}$. The refraction decreases from (a), where $a_2 = 0.25°$, over (b) with $a_2 = 0.23°$ to (c), where $a_2 = 0.18°$. The propagation direction is marked by the white arrow, whereas the interface is depicted by the dashed-dotted line.

The next peculiar feature of refraction between two discrete media is the fact that the ratio between transmission and reflection is not unique for a specific angle of incidence. In periodic media, the transverse component of the individual Bloch waves range within the first Brillouin zone from $\kappa_{min} = 0$ to $\kappa_{max} = \pi$, whereas the angle of propagation ranges only from $\alpha_{min} = 0°$ to $\alpha_{max} = \arctan\{-2\Delta c\}$, so that two different values of κ yield the same α.

Due to equations, in general the reflection and transmission is not symmetric around $\kappa = \frac{\pi}{2}$. Hence, for the same angle α, two different reflection and transmission coefficients exist for two different Bloch waves residing at different positions in the first Brillouin zone.

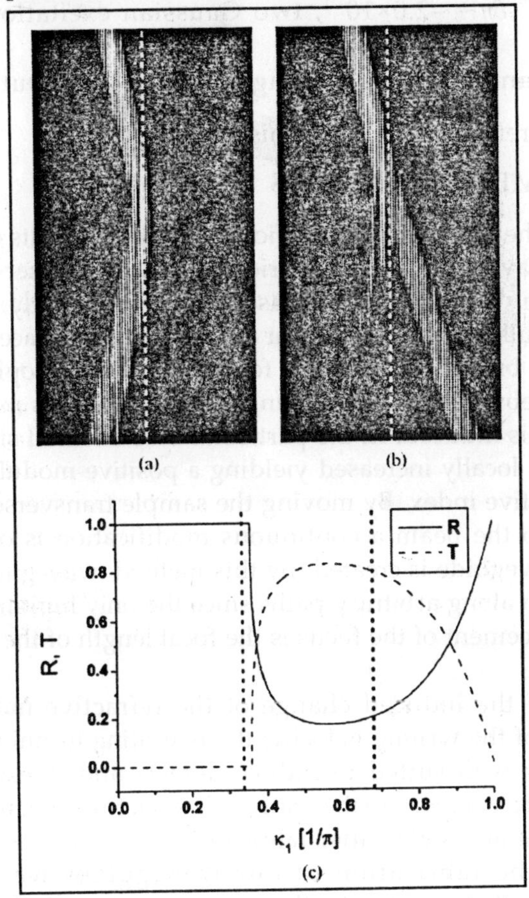

Fig. Partial reflection of a broad Gaussian excitation at an interface between two waveguide arrays with $c_1 = 1.05\ cm^{-1}$, $c_2 = 1.45\ cm^{-1}$ and a refractive index detuning of -2.0×10^{-4}.

In (a) the excitation is centered around $\kappa_1 = \frac{\pi}{3}$ and in (b) around $\kappa_1 = \frac{2\pi}{3}$, yielding in both cases a = 0.18°, but different reflection and transmission coefficients, which are depicted in (c). The propagation direction is from top to bottom.

This feature is depicted in figure. In an array with $c_1 = 1.05\ \text{cm}^{-1}$, $c_2 = 1.45\ \text{cm}^{-1}$ and a detuning of the refractive index of $\delta n = -2.0\times10^{-4}$, two Gaussian excitations with $\kappa_1^{(1)} = \frac{\pi}{3}$ and $\kappa_1^{(2)} = \frac{2\pi}{3}$ propagate at $\alpha = 0.18°$ but exhibit different reflection and transmission coefficients.

EXPERIMENTAL RESULTS

For the experimental verification of these results we used planar waveguide arrays fabricated by the fs laser writing technique. When ultrashort laser pulses are tightly focused into the bulk material, nonlinear absorption takes place leading to optical breakdown and the formation of a microplasma.

Thereby, a permanent change of the material's molecular structure is induced. In the particular case of fused silica, the density is locally increased yielding a positive modulation of the refractive index. By moving the sample transversely with respect to the beam a continuous modification is obtained and a waveguide is created. By this method, waveguides can be written along arbitrary paths since the only limiting factor in the placement of the focus is the focal length of the writing objective.

Since the induced change of the refractive index is a function of the writing velocity, the fs writing technique is in particular well suited to induce defects and interfaces in waveguide arrays. Furthermore, all structural changes are permanent and stable after fabrication.

For the fabrication of our waveguides we used a Ti:sapphire laser system with a repetition rate of 100 kHz, a pulse duration of about 150 fs and 0.3 μJ pulse energy at a laser wavelength of 800 nm. The beam was focused into polished fused-silica samples by a 20× microscope objective with a numerical aperture of 0.35.

The average power of the laser beam was 32 mW, while the writing velocity was 1300 μm s^{-1} resulting in a refractive index change of 6×10^{-4}. The length of all samples was 70 mm. To directly monitor the light propagation we applied a recently developed method of detecting fluorescent light.

For that purpose, we used a particular type of fused silica glass with a high content of OH which leads to a massive formation of non-bridging oxygen hole (NBOH) color centers in the modified volume. These can be excited with a HeNe source at 633 nm, which was launched by fibre butt coupling, so that the fluorescence with its maximum at 650 nm can be detected by a camera from above the sample .

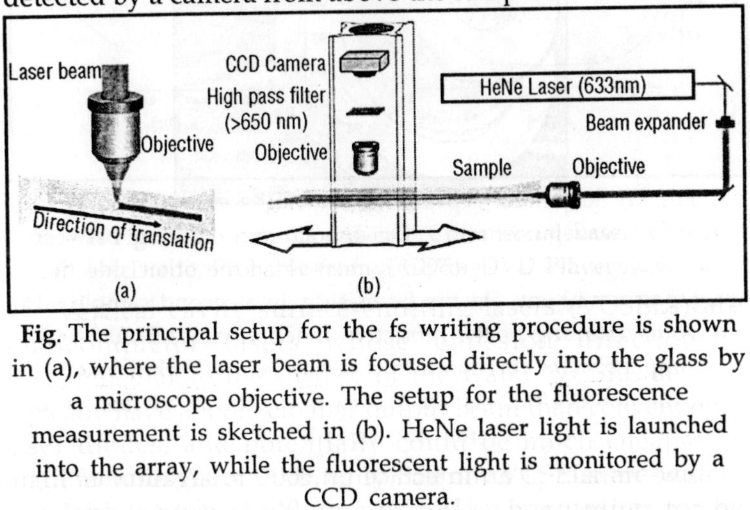

Fig. The principal setup for the fs writing procedure is shown in (a), where the laser beam is focused directly into the glass by a microscope objective. The setup for the fluorescence measurement is sketched in (b). HeNe laser light is launched into the array, while the fluorescent light is monitored by a CCD camera.

Two different setups were investigated experimentally, with a spacing of 13 μm and 15 μm on the left and right side, respectively. Additionally, in a second sample a detuning of the refractive index of $\delta n \approx -2\times10^{-4}$ was applied in the region with 13 μm spacing, which was achieved by a slightly higher writing speed of 1800 μm s^{-1}. The resulting coupling coefficient in the region with $\Delta = 15$ μm is $c_1 = 1.09$ cm^{-1}, in the region with $\Delta = 13$ μm is $c_2 = 1.56$ cm^{-1} .

These values remain valid even in the case of a slightly negative detuning, since the lower refractive index of the waveguides is compensated by a stronger overlap of adjacent modes. The respective calculated reflection and transmission coefficients of these setups are depicted in figures.

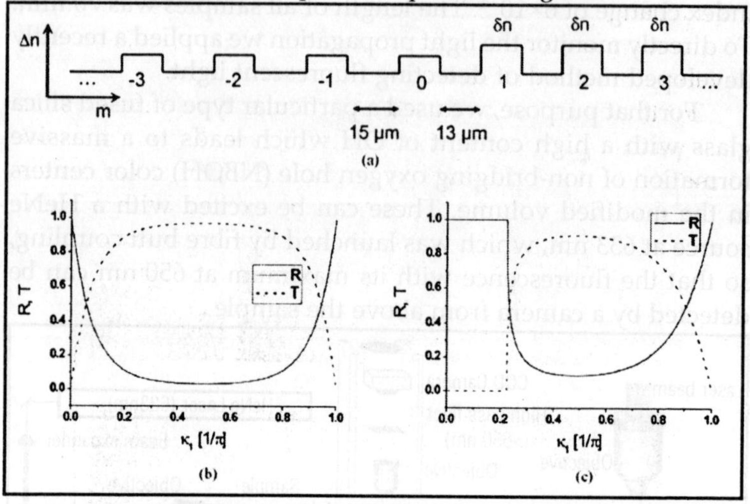

Fig. (a) Schematic representation of the experimentally analyzed waveguide array. On one side the spacing between the waveguides is $\Delta = 15$ μm, whereas on the other side the spacing is $\Delta = 13$ μm. On the latter side, in a second sample, also a small detuning of the refractive index of $\delta n \approx -1.5\times10^{-4}$ was applied. The corresponding calculated reflection and transmission coefficients are shown in (b) for $\delta n = 0$ and in (c) for $\delta n \approx -1.5\times10^{-4}$.

The impact of an interface on the propagation of light from a point source (exciting all Bloch waves) without additional detuning ($\delta n = 0$) is demonstrated in figure. The experiments illustrate the light evolution for the excitation of waveguide $m = -3$, $m = -2$, $m = -1$ and $m = 1$, shown in figures.

The corresponding simulations using equation are depicted in figures. Depending on the excited waveguide, the light is partially reflected at the interface. From figure it follows that in particular the side lobes at $\kappa = \frac{\pi}{2}$ are almost entirely transmitted and only slightly refracted from $\alpha_1 = 0.18°$ to $\alpha_1 = 0.23°$.

In contrast, the Bloch waves in the centre and the edge of

the Brillouin zone experience strong reflection. This is supported by the observation that a large fraction of the light between the side lobes, which results from these Bloch waves, remains in this semi-array where the excited waveguide is located.

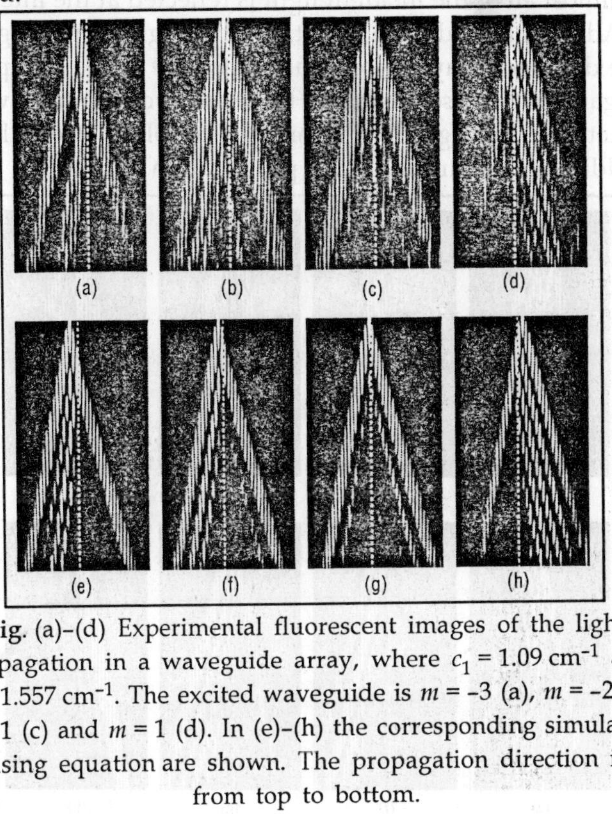

Fig. (a)–(d) Experimental fluorescent images of the light propagation in a waveguide array, where $c_1 = 1.09$ cm^{-1} and $c_2 = 1.557$ cm^{-1}. The excited waveguide is $m = -3$ (a), $m = -2$ (b), $m = -1$ (c) and $m = 1$ (d). In (e)–(h) the corresponding simulations using equation are shown. The propagation direction is from top to bottom.

This picture changes when an additional detuning $\delta n \approx -2\times10^{-4}$ is introduced in the ($m > 0$)-region, since in this case the reflection and transmission coefficients are modified.

The experimentally observed light evolution from a point source in such an array is depicted in figure for the excited waveguide at $m = 2$ and figure with the light launched in waveguide $m = 0$. The corresponding simulations are shown in figures.

The experimental propagation pattern of the undetuned

array is presented in figures for the excited waveguide at $m = 2$ and $m = 0$, respectively. The corresponding simulations are given in figures. The main feature is now that in the detuned array the reflection is much stronger for all Bloch waves, so that almost all of the incident light is reflected at the interface.

Additionally, due to equation, the angle of propagation in the detuned array region changes to $\alpha_1 = 0.20°$. This points out again that in discrete media there are two quantities which characterize the light evolution: the coupling strength and an additional detuning of the propagation constant.

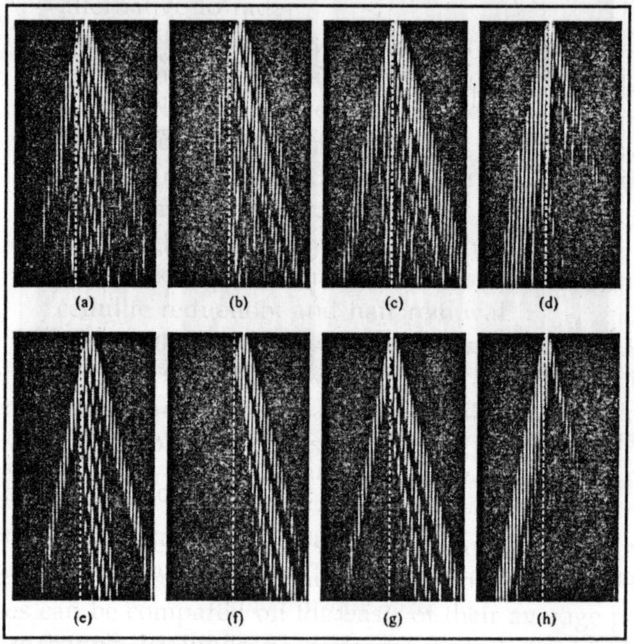

Fig. (a)–(d) Experimental fluorescent images of the light propagation in a waveguide array, where $c_1 = 1.09\ \text{cm}^{-1}$ and $c_2 = 1.557\ \text{cm}^{-1}$. In (a) and (b) the excited waveguide is $m = 2$, whereas in (c) and (d) this is $m = 0$. Furthermore, in (b) and (d) an additional detuning of $\delta n \approx -2\times10^{-4}$ is applied. The corresponding simulations are shown in (e)–(h). The propagation direction is from top to bottom.

We have presented the discrete analog of Snell's refraction

law at an interface. It was shown that the refraction is characterized by two quantities: the coupling strength and an additional detuning of the propagation constant, in contrast to continuous media. Additionally, refraction and transmission coefficients for the Bloch waves and the power were introduced as the discrete analog of Fresnel's formulae.

The results were experimentally verified in fs laser written waveguide arrays, where the light propagation was directly monitored using fluorescent light.

Chapter 10

Color and Vision

THE ELECTROMAGNETIC AND VISIBLE SPECTRA

Electromagnetic waves are waves which are capable of traveling through a vacuum. Unlike mechanical waves which require a medium in order to transport their energy, electromagnetic waves are capable of transporting energy through the vacuum of outer space. Electromagnetic waves are produced by a vibrating electric charge and as such, they consist of both an electric and a magnetic component.

Electromagnetic waves exist with an enormous range of frequencies. This continuous range of frequencies is known as the electromagnetic spectrum. The entire range of the spectrum is often broken into specific regions. The subdividing of the entire spectrum into smaller spectra is done mostly on the basis of how each region of electromagnetic waves interacts with matter. The diagram below depicts the electromagnetic spectrum and its various regions.

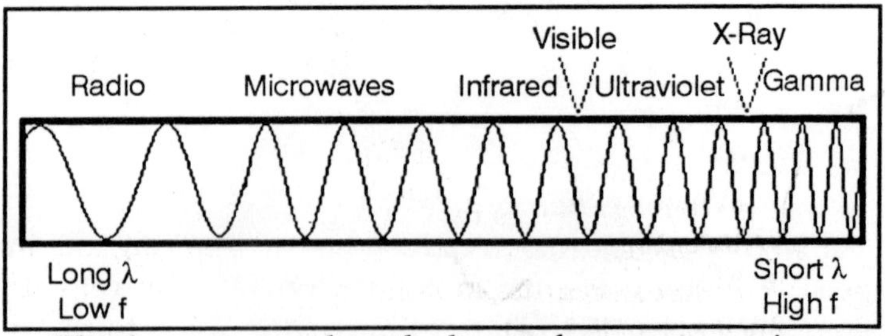

The longer wavelength, lower frequency regions are located on the far left of the spectrum and the shorter wavelength, higher frequency regions are on the far right.

Two very narrow regions within the spectrum are the visible light region and the X-ray region. We are undoubtedly familiar with some of the other regions of the electromagnetic spectrum.

Visible Light Spectrum

The visible light region - the very narrow band of wavelengths located to the right of the infrared region and to the left of the ultraviolet region. Though electromagnetic waves exist in a vast range of wavelengths, our eyes are sensitive to only a very narrow band. Since this narrow band of wavelengths is the means by which humans see, we refer to it as the visible light spectrum. Normally when we use the term "light," we are referring to a type of electromagnetic wave which stimulates the retina of our eyes.

In this sense, we are referring to visible light, a small spectrum from the enormous range of frequencies of electromagnetic radiation. This visible light region consists of a spectrum of wavelengths which range from approximately 700 nanometers (abbreviated nm) to approximately 400 nm. Expressed in more familiar units, the range of wavelengths extends from 7 x 10^{-7} meter to 4 x 10^{-7} meter. This narrow band of visible light is affectionately known as ROYGBIV.

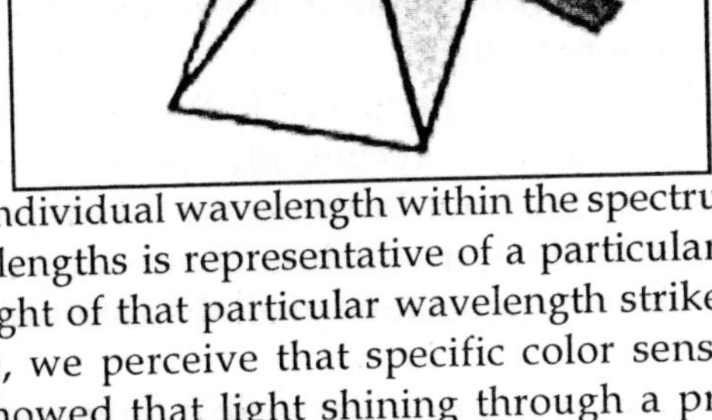

Each individual wavelength within the spectrum of visible light wavelengths is representative of a particular color. That is, when light of that particular wavelength strikes the retina of our eye, we perceive that specific color sensation. Isaac Newton showed that light shining through a prism will be separated into its different wavelengths and will thus show the various colors that visible light is comprised of. The separation of visible light into its different colors is known as dispersion.

Each color is characteristic of a distinct wavelength; and different wavelengths of light waves will bend varying amounts upon passage through a prism. For these reasons, visible light is dispersed upon passage through a prism. Dispersion of visible light produces the colors red (R), orange (O), yellow (Y), green (G), blue (B), and violet (V).

It is because of this that visible light is sometimes referred to as ROY G. BIV. (Incidentally, the indigo is not actually observed in the spectrum but is traditionally added to the list so that there is a vowel in Roy's last name.)

The red wavelengths of light are the longer wavelengths and the violet wavelengths of light are the shorter wavelengths. Between red and violet, there is a continuous range or spectrum of wavelengths. The visible light spectrum is shown in the diagram below.

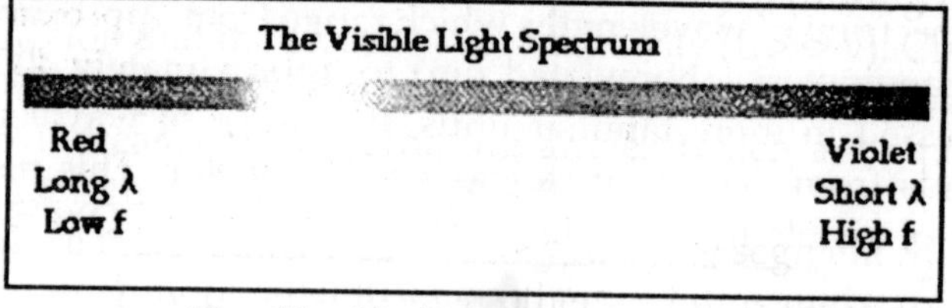

Fig. The Visible Light Spectrum

When all the wavelengths of the visible light spectrum strike our eye at the same time, white is perceived. The sensation of white is not the result of a single color of light. Rather, the sensation of white is the result of a mixture of two or more colors of light.

Thus, visible light - the mix of ROYGBIV - is sometimes referred to as white light. Technically speaking, white is not a color at all - at least not in the sense that there is a light wave with a wavelength which is characteristic of white. Rather, white is the combination of all the colors of the visible light spectrum.

If all the wavelengths of the visible light spectrum give the appearance of white, then none of the wavelengths would lead to the appearance of black. Once more, black is not actually a color. Technically speaking, black is merely the absence of the wavelengths of the visible light spectrum. So when we are in a room with no lights and everything around we appears black, it means that there are no wavelengths of visible light striking our eye as we sight at the surroundings.

Visible Light and the Eye's Response

Our eyes are sensitive to a very narrow band of frequencies within the enormous range of frequencies of the electromagnetic spectrum. This narrow band of frequencies is referred to as the visible light spectrum. Visible light - that which is detectable by the human eye - consists of wavelengths ranging from approximately 780 nanometer (7.80×10^{-7} m) down to 390 nanometer (3.90×10^{-7} m).

Specific wavelengths within the spectrum correspond to a specific color based upon how humans typically perceive light of that wavelength.

The long wavelength end of the spectrum corresponds to light which is perceived by humans to be red and the short wavelength end of the spectrum corresponds to light which is perceived to be violet. Other colors within the spectrum include orange, yellow, green and blue. The graphic below depicts the approximate range of wavelengths which are associated with the various perceived colors within the spectrum.

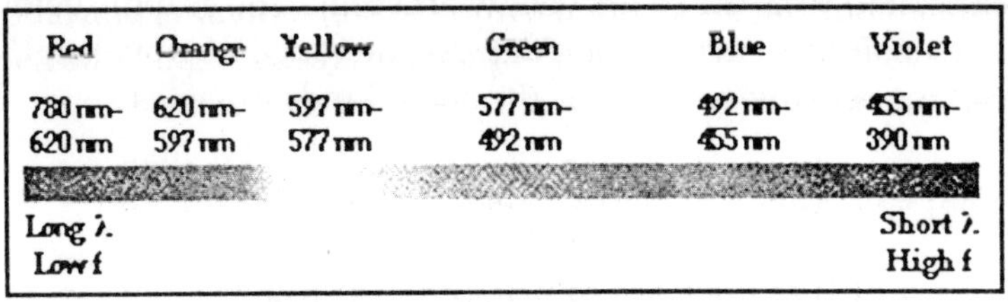

Fig. The Visible Light Spectrum

COLOR CONES

Color can be thought of as a psychological and physiological response to light waves of a specific frequency or set of frequencies impinging upon the eye. An understanding of the human response to color demands that one understand the biology of the eye. Light which enters the eye through the pupil ultimately strikes the inside surface of the eye known as the retina.

The retina is lined with a variety of light sensing cells known as rods and cones. While the rods on the retina are sensitive to the intensity of light, they cannot distinguish between light of different wavelengths. On the other hand, the cones are the color-sensing cells of the retina. When light of a given wavelength enters the eye and strikes the cones of the retina, a chemical reaction is activated which results in an electrical impulse being sent along nerves to the brain.

It is believed that there are three kinds of cones, each sensitive to its own range of wavelengths within the visible light spectrum. These three kinds of cones are referred to as red cones, green cones, and blue cones because of their respective sensitivity to the wavelengths of light which are associated with red, green and blue.

Since the red cone is sensitive to a range of wavelengths, it is not only activated by wavelengths of red light, but also (to a lesser extent) by wavelengths of orange light, yellow light and even green light.

In the same manner, the green cone is most sensitive to wavelengths of light associated with the color green. Yet the green cone can also be activated by wavelengths of light associated with the colors yellow and blue. The graphic below is a sensitivity curve which depicts the range of wavelengths and the sensitivity level for the three kinds of cones.

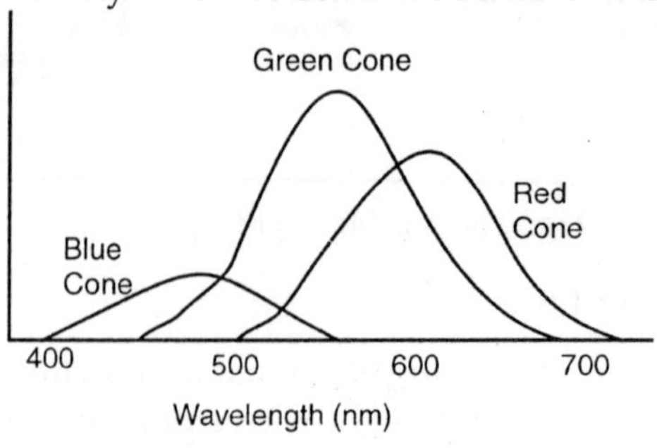

The cone sensitivity curve helps us to better understand our response to the light which is incident upon the retina. While the response is activated by the physics of light waves, the response itself is both physiological and psychological. Suppose that white light - i.e., light consisting of the full range of wavelengths within the visible light spectrum - is incident upon the retina. Upon striking the retina, the physiological occurs: photochemical reactions occur within the cones to produce electrical impulses which are sent along nerves to the brain.

The cones respond to the incident light by sending a message forward to brain, saying "light is hitting me." Upon reaching the brain, the psychological occurs: the brain detects the electrical messages being sent by the cones and interprets the meaning of the messages. The brain responds by saying "it is white."

For the case of white light entering the eye and striking the retina, each of the three kinds of cones would be activated into sending the electrical messages along to the brain. And the brain recognizes that the messages are being sent by all three cones and somehow interprets this to mean that white light has entered the eye.

Now suppose that light in the yellow range of wavelengths (approximately 577 nm to 597 nm) enters the eye and strikes the retina. Light with these wavelengths would activate both the green and the red cones of the retina. Upon striking the retina, the physiological occurs: electrical messages are sent by both the red and the green cones to the brain.

Once received by the brain, the psychological occurs: the brain recognizes that the light has activated both the red and the green cones and somehow interprets this to mean that the object is yellow. In this sense, the yellow appearance of objects is simply the result of yellow light from the object entering our eye and stimulating the red and the green cones simultaneously.

If the appearance of yellow is perceived of an object when it activates the red and the green cones simultaneously, then what appearance would result if two overlapping red and green spotlights entered our eye? Using the same *three-cone theory*, we could make some predictions of the result. Red light entering our eye would mostly activate the red color cone; and green light entering our eye would mostly activate the green color cone.

Each cone would send their usual electrical messages to the brain. If the brain has been psychologically trained to interpret these two signals to mean "yellow", then the brain would perceive the overlapping red and green spotlights to appear as yellow. To the eye-brain system, there is no difference in the physiological and psychological response to yellow light and a mixing of red and green light. The brain has no means of distinguishing between the two physical situations.

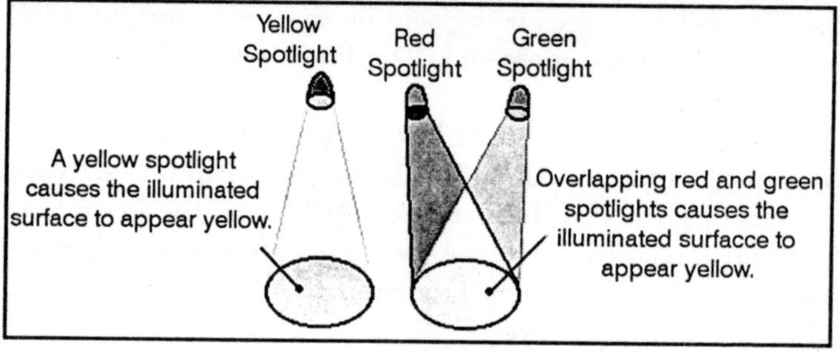

Fig. To the Eye-brain System, there is no Difference between Yellow Light and Overlapping Red and Green Light

In a technical sense, it is really not appropriate to refer to light as being colored. Light is simply a wave with a specific wavelength or a mixture of wavelengths; it has no color in and of itself. An object which is emitting or reflecting light to our eye appears to have a specific color as the result of the eye-brain response to the wavelength. So technically, there is really no such thing as yellow light. Rather, there is light with a wavelength of about 590 nm which appears yellow.

And there is also light with a mixture of wavelengths of about 700 nm and 530 nm which together appears yellow. The yellow appearance of these two clearly different light sources can be traced to the physiological and psychological response of the eye-brain system, and not to the light itself. So to be technically appropriate, a person would refer to "yellow light" as "light which creates a yellow appearance." Yet, to maintain a larger collection of friendships, a person would refer to "yellow light" as "yellow light."

Light Absorption, Reflection, and Transmission

We have previously learned that visible light waves consists of a continuous range of wavelengths or frequencies. When a light wave with a single frequency strikes an object, a number of things could happen. The light wave could be absorbed by the object, in which case its energy is converted to heat.

The light wave could be reflected by the object. And the light wave could be transmitted by the object. Rarely however does just a single frequency of light strike an object. While it does happen, it is more usual that visible light of many frequencies or even all frequencies are incident towards the surface of objects. When this occurs, objects have a tendency to selectively absorb, reflect or transmit light certain frequencies.

That is, one object might reflect green light while absorbing all other frequencies of visible light. Another object might selectively transmit blue light while absorbing all other frequencies of visible light. The manner in which visible light interacts with an object is dependent upon the frequency of the light and the nature of the atoms of the object.

Atoms and molecules contain electrons. It is often useful to think of these electrons as being attached to the atoms by springs. The electrons and their attached springs have a tendency to vibrate at specific frequencies. Similar to a tuning fork or even a musical instrument, the electrons of atoms have a natural frequency at which they tend to vibrate.

When a light wave with that same natural frequency impinges upon an atom, then the electrons of that atom will be set into vibrational motion. If a light wave of a given frequency strikes a material with electrons having the same vibrational frequencies, then those electrons will absorb the energy of the light wave and transform it into vibrational motion.

During its vibration, the electrons interacts with neighboring atoms in such a manner as to convert its vibrational energy into thermal energy. Subsequently, the light wave with that given frequency is absorbed by the object, never again to be released in the form of light.

So the selective absorption of light by a particular material occurs because the selected frequency of the light wave matches the frequency at which electrons in the atoms of that material vibrate. Since different atoms and molecules have different natural frequencies of vibration, they will selectively absorb different frequencies of visible light.

Reflection and transmission of light waves occur because the frequencies of the light waves do not match the natural frequencies of vibration of the objects. When light waves of these frequencies strike an object, the electrons in the atoms of the object begin vibrating. But instead of vibrating in resonance at a large amplitude, the electrons vibrate for brief periods of time with small amplitudes of vibration; then the energy is reemitted as a light wave.

If the object is transparent, then the vibrations of the electrons are passed on to neighboring atoms through the bulk of the material and reemitted on the opposite side of the object. Such frequencies of light waves are said to be transmitted. If the object is opaque, then the vibrations of the electrons are not passed from atom to atom through the bulk of the material. Rather the electrons of atoms on the material's surface vibrate for short periods of time and then reemit the energy as a reflected light wave. Such frequencies of light are said to be reflected.

The color of the objects which we see are largely due to the way those objects interact with light and ultimately reflect or transmit it to our eyes. The color of an object is not actually within the object itself. Rather, the color is in the light which shines upon it and is ultimately reflected or transmitted to our eyes. We know that the visible light spectrum consists of a range of frequencies, each of which corresponds to a specific color. When visible light strikes an object and a specific frequency becomes absorbed, that frequency of light will never make it to our eyes. Any visible light which strikes the object and becomes reflected or transmitted to our eyes will contribute to the color appearance of that object.

So the color is not in the object itself, but in the light which strikes the object and ultimately reaches our eye. The only role that the object plays is that it might contain atoms capable of selectively absorbing one or more frequencies of the visible light which shine upon it.

So if an object absorbs all of the frequencies of visible light except for the frequency associated with green light, then the object will appear green in the presence of ROYGBIV. And if an object absorbs all of the frequencies of visible light except for the frequency associated with blue light, then the object will appear blue in the presence of ROYGBIV.

COLOR ADDITION

Color perception, like sound perception, is a complex subject involving the disciplines of psychology, physiology, biology, chemistry and physics. When we look at an object and perceive a distinct color, we are not necessarily seeing a single frequency of light. Consider for instance that we are looking at a shirt and it appears purple to our eye. In such an instance, there may be several frequencies of light striking our eye with varying degrees of intensity. Yet our eye-brain system interprets the frequencies which strike our eye and the shirt is decoded by our brain as being *purple.*

The subject of color perception can be simplified if we think in terms of primary colors of light. We have already learned that white is not a color at all, but rather the presence of all the frequencies of visible light. When we speak of white light, we are referring to ROYGBIV - the presence of the entire spectrum of visible light. But combining the range of frequencies in the visible light spectrum is not the only means of producing white light.

White light can also be produced by combining only three distinct frequencies of light, provided that they are widely separated on the visible light spectrum. Any three colors (or frequencies) of light which produce white light when combined with the correct intensity are called primary colors of light. There are a variety of sets of primary colors. The most common set of primary colors is red (R), green (G) and blue (B). When red, green and blue light are mixed or added together with the proper intensity, white (W) light is obtained. This is often represented by the equation below:

$$R + G + B = W$$

In fact, the mixing together (or addition) of two or three of these three primary colors of light with varying degrees of intensity can produce a wide range of other colors. For this reason, many television sets and computer monitors produce the range of colors on the monitor by the use of of red, green and blue light-emitting phosphors.

The addition of the primary colors of light can be demonstrated using a light box. The light box illuminates a screen with the three primary colors - red (R), green (G) and blue (B). The lights are often the shape of circles. The result of adding two primary colors of light is easily seen by viewing the overlap of the two or more circles of primary light. The different combinations of colors produced by red, green and blue are shown in the graphic below.

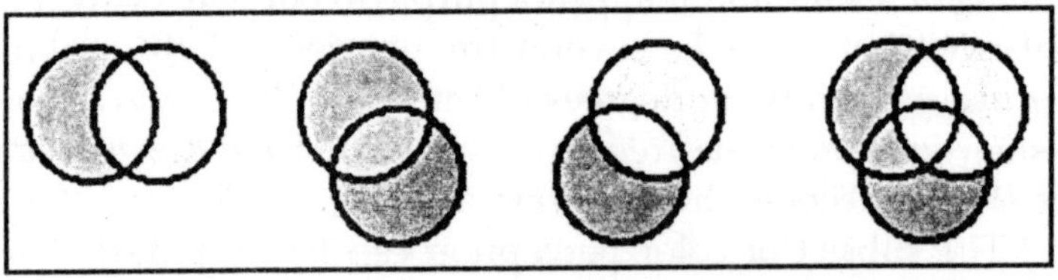

These demonstrations with the color box illustrate that red light and green light add together to produce yellow (Y) light. Red light and blue light add together to produce magenta (M) light.

Green light and blue light add together to produce cyan (C) light. And finally, red light and green light and blue light add together to produce white light. This is sometimes demonstrated by the following color equations and graphic:

Yellow (Y), magenta (M) and cyan (C) are sometimes referred to as secondary colors of light since they are produced by the addition of equal intensities of two primary colors of light. The addition of these three primary colors of light with varying degrees of intensity will result in the countless other colors which we are familiar (or unfamiliar) with.

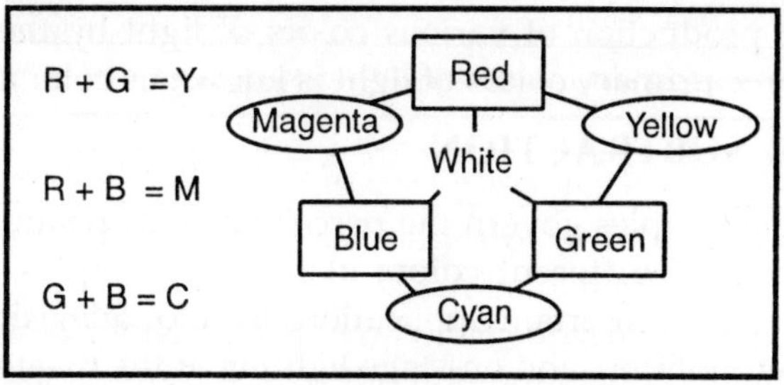

Complementary Colors of Light

Any two colors of light which when mixed together in equal intensities produce white are said to be complementary colors of each other. The complementary color of red light is cyan light.

This is reasonable since cyan light is the combination of blue and green light; and blue and green light when added to red light will produce white light. Thus, red light and cyan light (blue + green) represent a pair of complementary colors; they add together to produce white light. This is illustrated in the equation below:

$$R + C = R + (B + G) = \text{White}$$

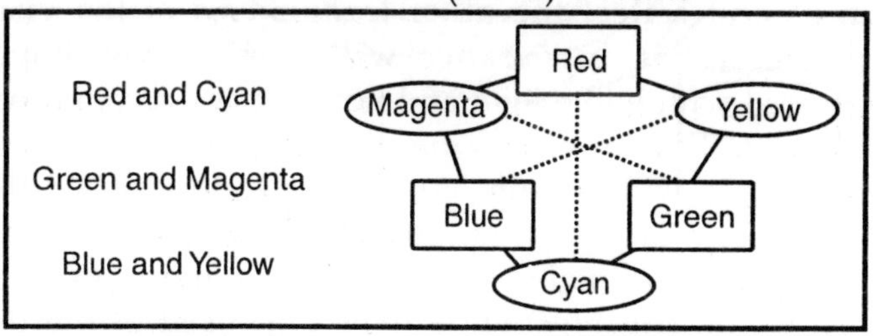

Fig. Complementary Colors of Light

Each primary color of light has a secondary color of light as its complement. The three pairs of complementary colors are listed below.

The graphic at the right is extremely helpful in identifying complementary colors. Complementary colors are always located directly across from each other on the graphic. Note that cyan is located across from red, magenta across from green, and yellow across from blue.

The production of various colors of light by the mixing of the three primary colors of light is known as color addition.

COLOR SUBTRACTION

The principles govern the perceived color resulting from the mixing of different colors of light. Principles of color addition have important applications to color television, color computer monitors and on-stage lighting at the theaters. Each of these applications involve the mixing or addition of colors of light to produce a desired appearance.

Our understanding of color perception would not be complete without an understanding of the principles of color subtraction. In this part we will learn how materials which have been permeated by specific pigments will selectively absorb specific frequencies of light in order to produce a desired appearance.

We have already learned that materials contain atoms which are capable of selectively absorbing one or more frequencies of light. Consider a shirt made of a material which is capable of absorbing blue light.

Such a material will absorb blue light (if blue light shines upon it) and reflect the other frequencies of the visible spectrum. What appearance will such a shirt have if illuminated with white light and how can we account for its appearance? To answer this question (and any other similar question), we will rely on our understanding of the three primary colors of light (red, green and blue) and the three secondary colors of light (magenta, yellow and cyan)

To begin, consider white light to consist of the three primary colors of light - red, green and blue. If white light is shining on a shirt, then red, green and blue light are shining on the shirt. If the shirt absorbs blue light, then only red and green light will be reflected from the shirt. So while red, green and blue light shine upon the shirt, only red and green light will reflect from it. Red and green light striking our eye always give the appearance of yellow; for this reason, the shirt will appear yellow.

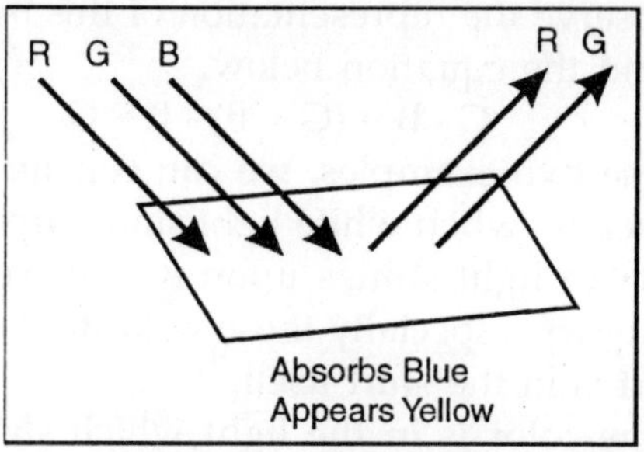

Fig. Color Substraction

This discussion illustrates the process of color subtraction. In this process, the ultimate color appearance of an object is determined by beginning with a single color or mixture of colors and identifying which color or colors of light are subtracted from the original set. The process is depicted visually by diagram at the right. Furthermore, the process is depicted in terms of an equation in the space below.

$$W - B = (R + G + B) - B = R + G = Y$$

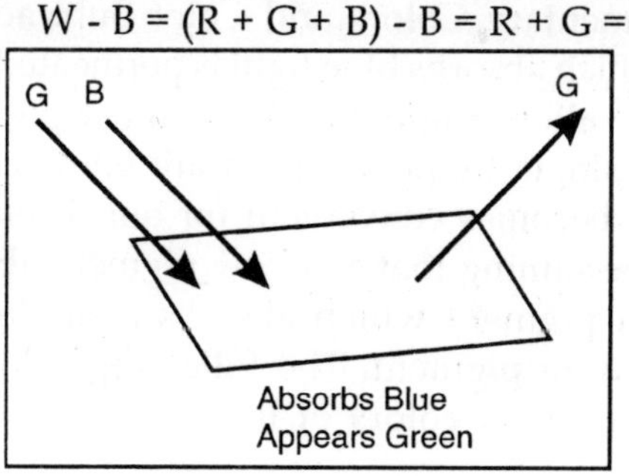

Fig. Color Substraction

Now suppose that cyan light is shining on the same shirt - a shirt made of a material which is capable of absorbing blue light. What appearance will such a shirt have if illuminated with cyan light and how can we account for its appearance? To answer this question, the process of color subtraction will be applied once more.

In this situation, we begin with only blue and green

primary colors of light. From this mixture, we must subtract blue light. After the subtractive process, only green light remains. Thus, the shirt will appear green in the presence of cyan light. Observe the representation of this by the diagram at the right and the equation below.

$$C - B = (G + B) - B = G$$

From these two examples, we can conclude that a shirt which looks yellow when white light shines upon it will look green when cyan light shines upon it. This confuses many students of physics, especially those who still believe that the color of a shirt is in the shirt itself.

Rather, the color is in the light which shines upon the object and which ultimately becomes reflected or transmitted to our eyes. Extending this conception of color, we would reason that the shirt appears yellow if there is some red and green light shining upon it. Yellow light is a combination of red and green light. A shirt appears yellow if it reflects red and green light to our eyes. In order to reflect red and green light, these two primary colors of light must be present in the incident light.

Complementary Colors and Color Subtraction

Paper which absorbs blue light is permeated by a pigment known as a yellow pigment. While most pigments absorb more than a single frequency (and are known as compound pigments), it becomes convenient for our discussion to keep it simple by assuming that a yellow pigment absorbs a single frequency. A pigment which absorbs a single frequency is known as a pure pigment. The following rule will assist in understanding what colors of light are absorbed by which pigments.

Pigments absorb light. Pure pigments absorb a single frequency or color of light. The color of light absorbed by a pigment is merely the complementary color of that pigment.

Thus, pure blue pigments absorb yellow light (which can be thought of as a combination of red and green light). Pure yellow pigments absorb blue light. Pure green pigments absorb magenta light (which can be thought of as a combination of red and blue light). Pure magenta pigments absorb green light. Pure red pigments absorb cyan light (which can be thought of as a combination of blue and green light). And finally, pure cyan pigments absorb red light.

Now lets combine the process of color subtraction with an understanding of complementary colors to determine the color appearance of various sheets of paper when illuminated by various lights. We will investigate three examples.

Example 1

Magenta light shines on a sheet of paper containing a yellow pigment. Determine the appearance of the paper.

Magenta light can be thought of as consisting of red light and blue light. A yellow pigment is capable of absorbing blue light. Thus, blue is subtracted from the light which shines on the paper. This leaves red light. If the paper reflects the red light, then the paper will look red.

$$M - B = (R + B) - B = R$$

Example 2

Yellow light shines on a sheet of paper containing a red pigment. Determine the appearance of the paper.

Yellow light can be thought of as consisting of red light and green light. A red pigment is capable of absorbing cyan light. That is, red paper can absorb both green and blue primary colors of light. So red and green light shine on the paper; and green light is subtracted. (There is no need to subtract blue light since blue light is not shining on the paper.) This leaves red light to be reflected. If the paper reflects the red light, then the paper will look red.

$$Y - G = (R + G) - G = R$$

Example 3

Yellow light shines on a sheet of paper containing a blue pigment. Determine the appearance of the paper.

Yellow light can be thought of as consisting of red light and green light. A blue pigment is capable of absorbing yellow light. That is, blue paper can absorb both red and green primary colors of light (recall that yellow light is a mixture of red and green light). So red and green light shine on the paper; and both the red and the green light are subtracted. There is no color left to be reflected to the eye. Subsequently, the paper appears black.

Y - Y = (R + G) - (R + G) = No reflected light = Black

FILTERS AND COLOR SUBTRACTION

The distinction between opaque and transparent materials was made earlier in this lesson. Opaque materials selectively absorb one or more frequencies of light and reflect what is not absorbed. In contrast to opaque materials, transparent materials selectively absorb one or more frequencies of light and transmit what is not absorbed.

Like opaque materials, transparent materials are permeated by pigments which contain atoms that are capable of absorbing light with a single frequency or even a range of frequencies. Knowing the color(s) of the incident light and the color of light absorbed by the pigment or filter, the process of color subtraction can be applied to determine the color appearance of a transparent material. We will consider three examples in the space below; the examples are visually depicted in the diagrams below.

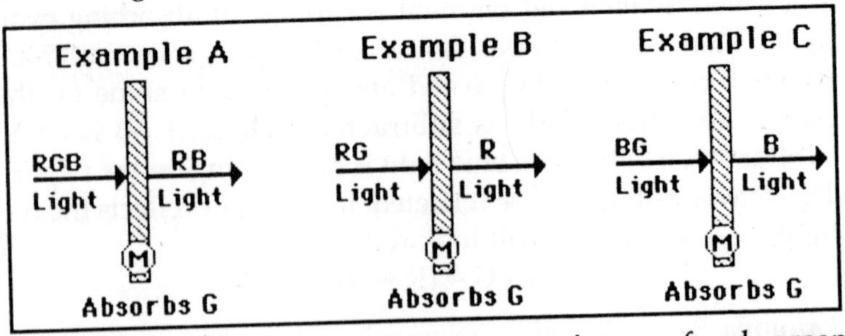

In Example A, white light (i.e., a mixture of red, green and blue) shines upon a magenta filter. Magenta absorbs its complementary color - green.

Thus, green is subtracted from white light. That leaves red and blue light to be transmitted by the filter. For this reason, the filter will appear magenta (recall that magenta light is a mixture of red and blue light) when illuminated with white light. This process of color subtraction can be represented by the following equation.

$$W - G = (R + G + B) - G = R + B = M$$

In Example B, yellow light (i.e., a mixture of red and green) shines upon the same magenta filter. Magenta absorbs its complementary color - green. Thus, green is subtracted from yellow light. That leaves red light to be transmitted by the filter. For this reason, the filter will appear red when illuminated with yellow light. This process of color subtraction can be represented by the following equation.

$$Y - G = (R + G) - G = R$$

In Example C, cyan light (i.e., a mixture of blue and green) shines upon the same magenta filter. Magenta absorbs its complementary color - green. Thus, green is subtracted from cyan light. That leaves blue light to be transmitted by the filter. For this reason, the filter will appear blue when illuminated with cyan light. This process of color subtraction can be represented by the following equation.

$$C - G = (B + G) - G = B$$

The reasoning modeled in the above three examples can be used in any situation, regardless of the color of the incident light and the color of the filter. As we approach such problems, whether they involve transparent or opaque materials, be sure to think in terms of primary colors of light and to use the logical reasoning steps. Avoid memorizing and avoid shortcuts. If a filter is capable of absorbing a color of light which is not present in the mixture of incident light, then merely disregard that color. Since that color of light is not incident upon the object, it cannot contribute to the color appearance of the object.

Primary Colors of Paint

A trip to the local newspaper or film developing company will reveal these same principles of color subtraction at work. The three primary colors of paint used by an artist, color printer or film developer are cyan (C), magenta (M), and yellow (Y). Artists, printers, and film developers do not deal directly with light; rather, they must apply paints or dyes to a white sheet of paper. These paints and dyes must be capable of absorbing the appropriate components of white light in order to produce the desired affect.

Most artists start with a white canvas and apply paints. These paints have to subtract colors so that we might see the desired image. An artist can create any color by using varying amounts of these three primary colors of paint.

Each primary color of paint absorbs one primary color of light. The color absorbed by a primary color of paint is the complementary color of that paint. Each of the three colors which are primary to an artist (magenta, cyan, and yellow) subtract red, green, and blue individually from an otherwise white sheet of paper. Thus,

Magenta paints absorb green light.

Cyan paints absorb red light.

Yellow paints absorb blue light.

Let's suppose that an artist wishes to use the three primary colors of paint in order to produce a picture of the colorful bird shown at the right. The bird will be painted onto white paper and viewed under white light. It is hoped that the bird will have green tail feathers, a blue lower body, a cyan upper body, a red head, a magenta eye patch, a yellow eye and middle feathers, and a black beak. How can the three primary colors of paint be used to produce such a likeness? And how can we explain the answers in terms of color subtraction?

To produce a green tail, paints must be applied to the tail region in order to absorb red and blue light and leave green to be reflected. Thus, the green tail must be painted using yellow paint (to absorb the blue) and cyan paint (to absorb the red).

To produce a blue lower body, paints must be applied to the lower body region in order to absorb red and green light, leaving blue light to be reflected. Thus, the blue lower body must be painted using magenta paint (to absorb the green) and cyan paint (to absorb the red).

To produce a red head, paints must be applied to the head region in order to absorb blue and green light, leaving red light to be reflected. Thus, the red head must be painted using magenta paint (to absorb the green) and yellow paint (to absorb the blue).

To produce a cyan upper body, paints must be applied to the upper body region in order to absorb red, leaving green and blue light to be reflected. If green and blue light are reflected from the upper body region, it will appear cyan (recall that blue and green light combine to form cyan light). Thus, the cyan upper body must be painted using merely cyan paint (to absorb the red).

To produce a magenta eye patch, paints must be applied to the eye patch region in order to absorb green, leaving red and blue light to be reflected. If red and blue light are reflected from the eye patch region, it will appear magenta (recall that blue and red light combine to form magenta light). Thus, the magenta eye patch must be painted using merely magenta paint (to absorb the green).

To produce a yellow eye and middle feathers, paints must be applied to the eye and middle feather regions in order to absorb blue, leaving red and green light to be reflected. If red and green light are reflected from the eye and middle feather regions, it will appear yellow (recall that red and green light combine to form yellow light). Thus, the yellow eye and middle feathers must be painted using merely yellow paint (to absorb the blue).

This information is summarized in the graphic below.

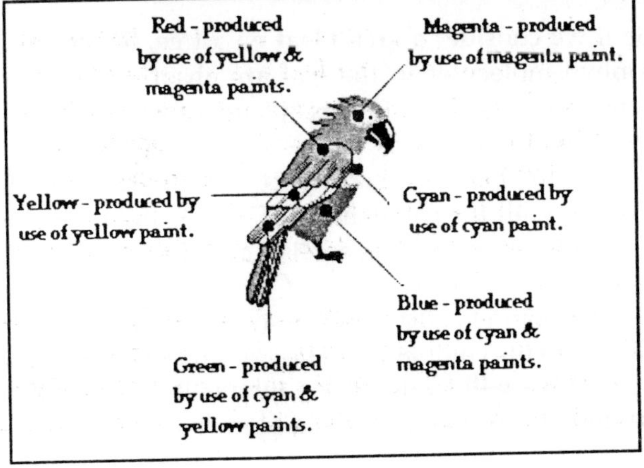

The process of color subtraction is a useful means of predicting the ultimate color appearance of an object if the color of the incident light and the pigments are known. By using the complementary color scheme, the colors of light which will be absorbed by a given material can be determined. These colors are subtracted from the incident light colors (if present) and the colors of reflected light (or transmitted light) can be determined. Then the color appearance of the object can be predicted.

Blue Skies and Red Sunsets

The sun emits light waves with a range of frequencies. Some of these frequencies fall within the visible light spectrum and thus are detectable by the human eye. Since sunlight consists of light with the range of visible light frequencies, it appears white. This white light is incident towards Earth and illuminates both our outdoor world and the atmosphere which surrounds our planet. The interaction of visible light with matter will often result in the absorption of specific frequencies of light. The frequencies of visible light which are not absorbed are either transmitted (by transparent materials) or reflected (by opaque materials).

As we sight at various objects in our surroundings, the color which we perceive is dependent upon the color(s) of light which are reflected or transmitted by those objects to our eyes.

So if we consider a green leaf on a tree, the atoms of the chlorophyll molecules in the leaf are absorbing most of the frequencies of visible light (except for green) and reflecting the green light to our eyes. The leaf thus appears green. And as we view the black asphalt street, the atoms of the asphalt are absorbing all the frequencies of visible light and no light is reflected to our eyes. The asphalt street thus appears black (the absence of color).

In this manner, the interaction of sunlight with matter contributes to the color appearance of our surrounding world. In this part we will focus on the interaction of sunlight with atmospheric particles to produce blue skies and red sunsets.

We will attempt to answer these two questions:

- Why are the skies blue?
- Why are the sunsets red?

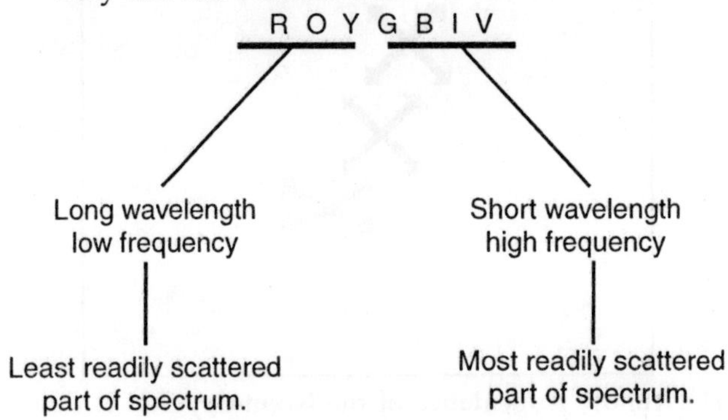

Fig. Visible Light Spectrum

The interaction of sunlight with matter can result in one of three wave behaviors: absorption, transmission, and reflection. The atmosphere is a gaseous sea which contains a variety of types of particles; the two most common types of matter present in the atmosphere are gaseous nitrogen and oxygen. These particles are most effective in scattering the higher frequency and shorter wavelength portions of the visible light spectrum. This scattering process involves the absorption of a light wave by an atom followed by reemission of a light wave in a variety of directions. The amount of multidirectional scattering which occurs is dependent upon the frequency of the light. (In fact, it varies according to f^4.)

Atmospheric nitrogen and oxygen scatter violet light most easily, followed by blue light, green light, etc. So as white light (ROYGBIV) from the sun passes through our atmosphere, the high frequencies (BIV) become scattered by atmospheric particles while the lower frequencies (ROY) are most likely to pass through the atmosphere without a significant alteration in their direction.

This scattering of the higher frequencies of light illuminates the skies with light on the BIV end of the visible spectrum. Compared to blue light, violet light is most easily scattered by atmospheric particles. However, our eyes are more sensitive to light with blue frequencies. Thus, we view the skies as being blue in color.

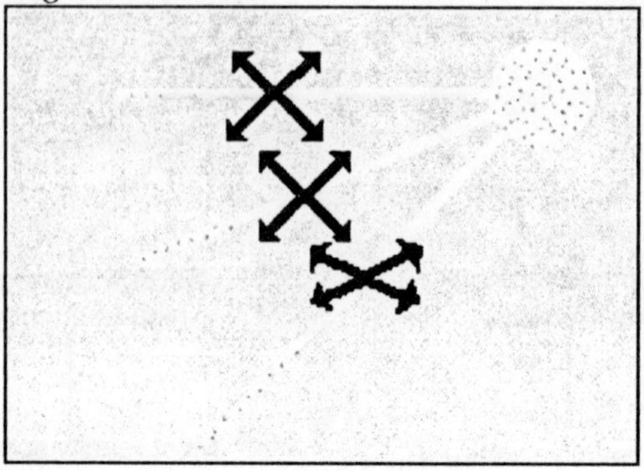

Fig. The Yellow Appeatance of the Noon-day Sun is due to the Scattering of the Higher Frequencies of Sunlight.

Meanwhile, the light that is not scattered is able to pass through our atmosphere and reach our eyes in a rather non-interrupted path. The lower frequencies of sunlight (ROY) tend to reach our eyes as we sight directly at the sun during midday.

While sunlight consists of the entire range of frequencies of visible light, not all frequencies are equally intense. In fact, sunlight tends to be most rich with yellow light frequencies. For these reasons, the sun appears yellow during midday due to the direct passage of dominant amounts of yellow frequencies through our atmosphere and to our eyes.

The appearance of the sun changes with the time of day. While it may be yellow during midday, it is often found to gradually turn color as it approaches sunset.

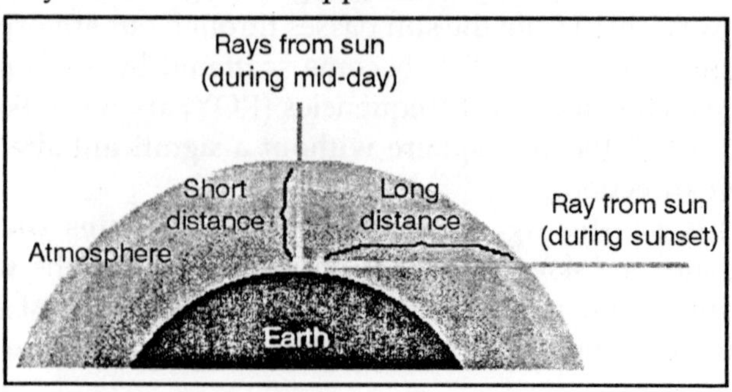

This can be explained by light scattering. As the sun approaches the horizon line, sunlight must traverse a greater distance through our atmosphere; this is demonstrated in the diagram below.

As the path which sunlight takes through our atmosphere increases in length, ROYGBIV encounters more and more atmospheric particles. This results in the scattering of greater and greater amounts of yellow light. During sunset hours, the light passing through our atmosphere to our eyes tends to be most concentrated with red and orange frequencies of light.

For this reason, the sunsets have a reddish-orange hue. The affect of a red sunset becomes more pronounced if the atmosphere contains more and more particles. The presence of sulfur aerosols (emitted as an industrial pollutant and by volcanic activity) in our atmosphere contributes to some magnificent sunsets (and some very serious environmental problems).

Chapter 11

Geometrical Optics

When a plane wave is impingent on an aperture which has dimensions much greater than the wavelength of the wave, diffraction effects are minimal and a segment of the plane wave passes through the aperture essentially unaltered. This plane wave segment can be thought of as a wave packet, called a beam or ray, consisting of a superposition of wave vectors very close in direction and magnitude to the central wave vector of the wave packet.

In most cases the ray simply moves in the direction defined by the central wave vector, i. e., normal to the orientation of the wave fronts. However, this is not true when the medium through which the light propagates is optically anisotropic, i. e., light traveling in different directions moves at different phase speeds. An example of such a medium is a calcite crystal.

If light moves through some apparatus in which all apertures are much greater in dimension than the wavelength of light, then we can use the above rule to follow rays of light through the apparatus. This is called the *geometrical optics* approximation.

REFLECTION AND REFRACTION

Most of what we need to know about geometrical optics can be summarized in two rules, the laws of reflection and refraction. These rules may both be inferred by considering what happens when a plane wave segment impinges on a flat surface.

If the surface is polished metal, the wave is *reflected*, whereas if the surface is an interface between two transparent media with differing indices of refraction, the wave is partially reflected and partially *refracted.*

Reflection means that the wave is turned back into the half-space from which it came, while refraction means that it passes through the interface, acquiring a different direction of motion from that which it had before reaching the interface.

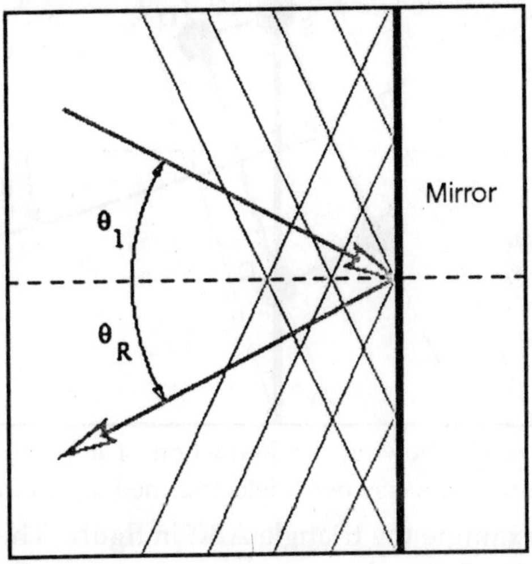

Fig. Sketch Showing the Reflection of a Wave from a Plane Mirror. The Law of Reflection States that $\theta_I = \theta_R$.

Figure shows the wave vector and wave front of a wave being reflected from a plane mirror. The angles of incidence, θ_I, and reflection, θ_R, are defined to be the angles between the incoming and outgoing wave vectors respectively and the line normal to the mirror. The law of reflection states that $\theta_R = \theta_I$. This is a consequence of the need for the incoming and outgoing wave fronts to be in phase with each other all along the mirror surface. This plus the equality of the incoming and outgoing wavelengths is sufficient to insure the above result.

Refraction, is slightly more complicated. Since $n_R > n_I$, the speed of light in the right-hand medium is less than in the left-hand medium. (Recall that the speed of light in a medium with refractive index n is $c_{medium} = c_{vac}/n$.) The frequency of the wave packet doesn't change as it passes through the interface, so the wavelength of the light on the right side is less than the wavelength on the left side.

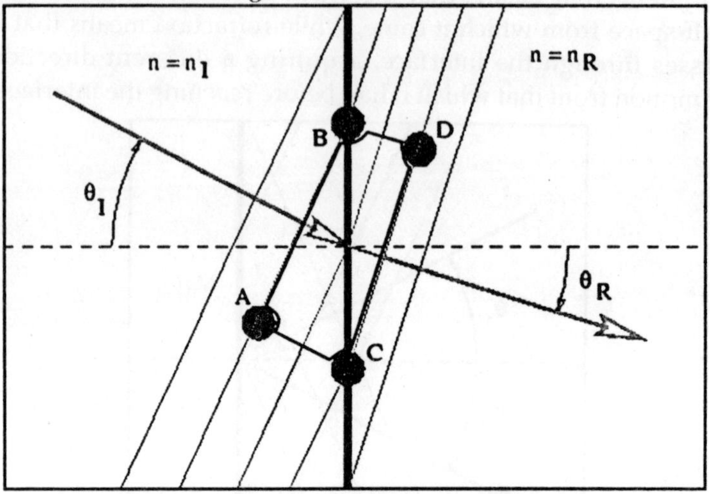

Fig. Sketch Showing the Refraction of a Wave from an Interface between two dielectric media with $n_2 > n_1$.

Let us examine the triangle ABC in figure. The side AC is equal to the side BC times $\sin(\theta_I)$. However, AC is also equal to $2\lambda_I$, or twice the wavelength of the wave to the left of the interface. Similar reasoning shows that $2\lambda_R$, twice the wavelength to the right of the interface, equals BC times. Since the interval BC is common to both triangles, we easily see that

$$\frac{\lambda_I}{\lambda_R} = \frac{\sin(\theta_I)}{\sin(\theta_R)}.$$

Since $\lambda_I = c_I T = c_{vac} T / n_I$ and $\lambda_R = c_R T = c_{vac} T / n_R$ where $C_I$and C_R are the wave speeds to the left and right of the interface, c_{vac} is the speed of light in a vacuum, and T is the (common) period, we can easily recast the above equation in the form

$$n_I \sin(\theta_I) = n_R \sin(\theta_R).$$

This is called *Snell's law*, and it governs how a ray of

light bends as it passes through a discontinuity in the index of refraction. The angle θ_I is called the incident angle and θ_R is called the refracted angle. Notice that these angles are measured from the normal to the surface, not the tangent.

TOTAL INTERNAL REFLECTION

When light passes from a medium of lesser index of refraction to one with greater index of refraction, Snell's law indicates that the ray bends *toward* the normal to the interface. The reverse occurs when the passage is in the other direction.

In this latter circumstance a special situation arises when Snell's law predicts a value for the sine of the refracted angle greater than one.

This is physically untenable. What actually happens is that the incident wave is *reflected* from the interface. This phenomenon is called *total internal reflection*. The minimum incident angle for which total internal reflection occurs is obtained by substituting $\theta_R = \pi/2$ into equation, resulting in

$$\sin(\theta_I) = n_R / n_I \text{ (total internal reflection).}$$

Anisotropic Media

Notice that Snell's law makes the implicit assumption that rays of light move in the direction of the light's wave vector, i. e., normal to the wave fronts. As the analysis makes clear, this is valid only when the optical medium is isotropic, i. e., the wave frequency depends only on the magnitude of the wave vector, not on its direction.

Certain kinds of crystals, such as those made of calcite, are not isotropic – the speed of light in such crystals, and hence the wave frequency, depends on the orientation of the wave vector. As an example, the angular frequency in an isotropic medium might take the form

$$\omega = \left[\frac{c_1^2\left(k_x + k_y\right)^2}{2} + \frac{c_2^2\left(k_x - k_y\right)^2}{2}\right]^{1/2},$$

where c_1 is the speed of light for waves in which $k_y = k_x$, and c_2 is its speed when $k_y = -k_x$.

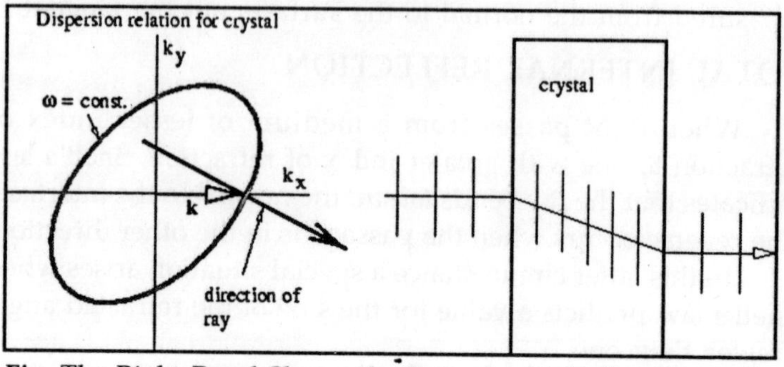

Fig. The Right Panel Shows the Fate of a Light Ray Normally Incident on the Face of a Properly Cut Calcite Crystal. The Anisotropic Dispersion Relation which Gives Rise to this Behaviour is Shown in the Left Panel.

Figure shows an example in which a ray hits a calcite crystal oriented so that constant frequency contours are as specified in equation. The wave vector is oriented normal to the surface of the crystal, so that wave fronts are parallel to this surface.

Upon entering the crystal, the wave front orientation must stay the same to preserve phase continuity at the surface. However, due to the anisotropy of the dispersion relation for light in the crystal, the ray direction changes as shown in the right panel. This behaviour is clearly inconsistent with the usual version of Snell's law!

It is possible to extend Snell's law to the anisotropic case. However, we will not present this here. The following discussions of optical instruments will always assume that isotropic optical media are used.

Thin Lens Formula and Optical Instruments

Given the laws of reflection and refraction, one can see in principle how the passage of light through an optical instrument could be traced. For each of a number of initial rays, the change in the direction of the ray at each mirror surface or refractive index interface can be calculated. Between these points, the ray traces out a straight line.

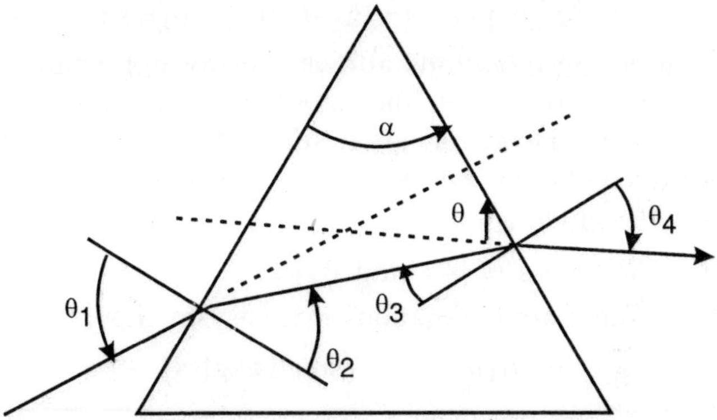

Fig. Bending of a Ray of Light as it Passes Through a Prism.

Though simple in conception, this procedure can be quite complex in practice.

However, the procedure simplifies if a number of approximations, collectively called the *thin lens approximation*, are valid. We begin with the calculation of the bending of a ray of light as it passes through a prism.

The pieces of information needed to find θ, the angle through which the ray is deflected are as follows: The geometry of the triangle defined by the entry and exit points of the ray and the upper vertex of the prism leads to

$$\alpha + (\pi/2 - \theta_2) + (\pi/2 - \theta_3) = -\pi,$$

which simplifies to

$$\alpha = \theta_2 + \theta_3.$$

Snell's law at the entrance and exit points of the ray tell us that

$$n = \frac{\sin(\theta_1)}{\sin(\theta_2)} \qquad n = \frac{\sin(\theta_4)}{\sin(\theta_3)},$$

where is the index of refraction of the prism. (The index of refraction of the surroundings is assumed to be unity.) One can also infer that

$$\theta = \theta_1 + \theta_4 - \alpha.$$

This comes from the fact that the the sum of the internal angles of the shaded quadrangle in figure is.

$$(\pi/2-\theta_1)+\alpha+(\pi/2-\theta_4)+(\pi+\theta)=2\pi$$

Combining equations allows the ray deflection to be determined in terms of θ_1 and α, but the resulting expression is very messy. However, great simplification occurs if the following conditions are met:

- The angle $\alpha \ll 1$.
- The angles $\theta_1 \ll 1$ and $\theta_4 \ll 1$.

With these approximations it is easy to show that

$$\theta=\alpha(n-1) \qquad \text{(small angles)}.$$

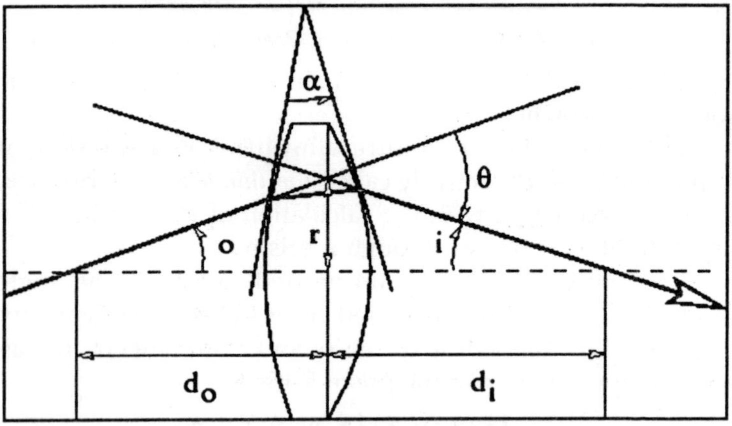

Fig. Light ray undergoing deflection through an angle θ by a lens. The angle α is the angle between the tangents to the entry and exit points of the ray on the lens.

Generally speaking, lenses and mirrors in optical instruments have curved rather than flat surfaces. However, we can still use the laws for reflection and refraction by plane surfaces as long as the segment of the surface on which the wave packet impinges is not curved very much on the scale of the wave packet dimensions.

This condition is easy to satisfy with light impinging on ordinary optical instruments. In this case, the deflection of a ray of light is given by equation if α is defined as the intersection of the tangent lines to the entry and exit points of the ray.

A *positive lens* is thicker in the centre than at the edges. The angle α between the tangent lines to the two surfaces of the lens at a distance r from the central axis takes the form $\alpha = Cr$, where C is a constant. The deflection angle of a beam hitting the lens a distance from the centre is therefore $\theta = Cr(n-1)$, as indicated in figure.

The angles o and i sum to the deflection angle: $o + i = \theta = Cr(n-1)$. However, to the extent that the small angle approximation holds $o = r/d_o$, and $i = r/d_i$ where d_o is the distance to the object and d_i is the distance to the image of the object. Putting these equations together and cancelling the r results in the *thin lens formula*:

$$\frac{1}{d_o} + \frac{1}{d_i} = C(n-1) \equiv \frac{1}{f}.$$

The quantity f is called the *focal length* of the lens. Notice that $f = d_i$ if the object is very far from the lens, i. e., if d_o is extremely large.

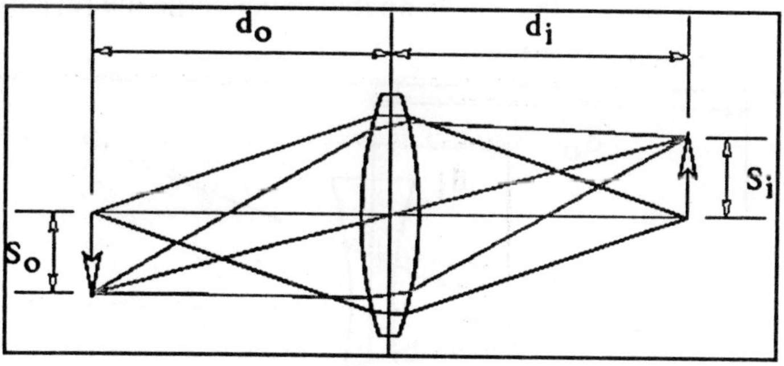

Fig. A Positive Lens Producing an Image on the Right of the Arrow on the Left.

Figure shows how a positive lens makes an image. The image is produced by all of the light from each point on the object falling on a corresponding point in the image. If the arrow on the left is an illuminated object, an *image* of the arrow will appear at right if the light coming from the lens is allowed to fall on a piece of paper or a ground glass screen. The size of the object S_0 and the size of the image S_i are related by simple geometry to the distances of the object and the image from the lens:

$$\frac{S_i}{S_o} = \frac{d_i}{d_o}.$$

Notice that a positive lens inverts the image.

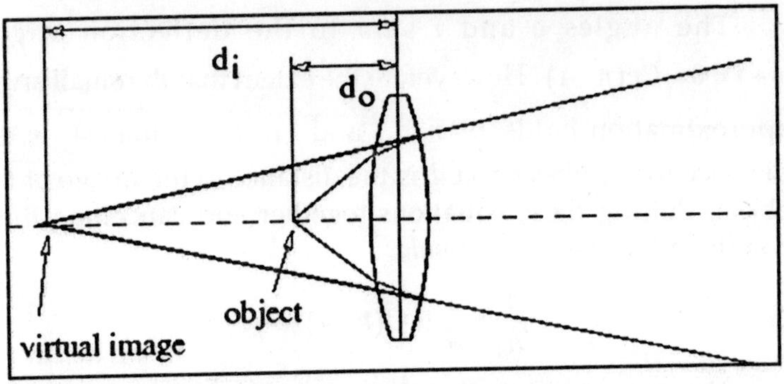

Fig. Production of a Virtual Image by a Positive Lens.

An image will be produced to the right of the lens only if $d_o > f$. If $d_o < f$, the lens is unable to converge the rays from the image to a point.

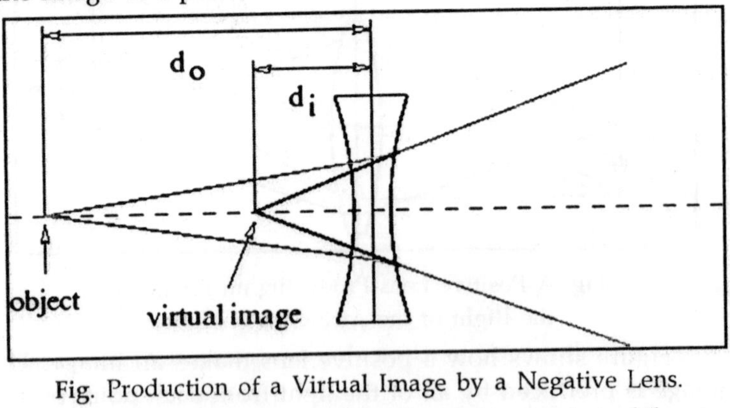

Fig. Production of a Virtual Image by a Negative Lens.

However, in this case the backward extension of the rays converge at a point called a *virtual image*, which in the case of a positive lens is always farther away from the lens than the object.

The thin lens formula still applies if the distance from the lens to the image is taken to be negative. The image is called virtual because it does not appear on a ground glass screen placed at this point. Unlike the real image seen in figure, the virtual image is not inverted.

A *negative lens* is thinner in the centre than at the edges and produces only virtual images. As seen in figure, the virtual image produced by a negative lens is closer to the lens than is the object. Again, the thin lens formula is still valid, but both the distance from the image to the lens and the focal length must be taken as negative. Only the distance to the object remains positive.

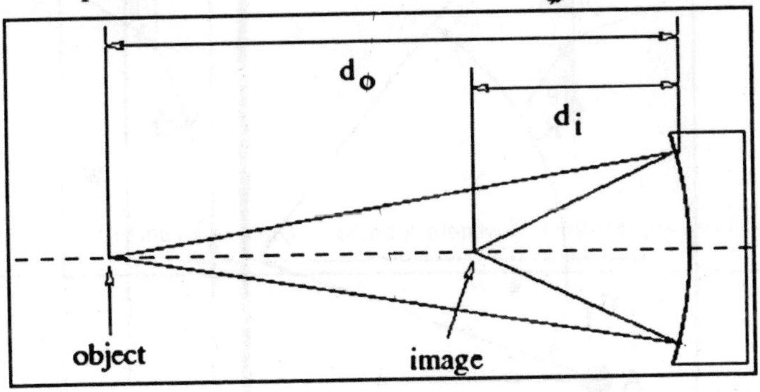

Fig. Production of a Real Image by a Concave Mirror.

Curved mirrors also produce images in a manner similar to a lens. A concave mirror, as seen in this figure, works in analogy to a positive lens, producing a real or a virtual image depending on whether the object is farther from or closer to the mirror than the mirror's focal length.

A convex mirror acts like a negative lens, always producing a virtual image. The thin lens formula works in both cases as long as the angles are small.

FERMAT'S PRINCIPLE

An alternate approach to geometrical optics can be developed from Fermat's principle. This principle states (in it's simplest form) that light waves of a given frequency traverse the path between two points which takes the least time.

The most obvious example of this is the passage of light through a homogeneous medium in which the speed of light doesn't change with position. In this case shortest time is equivalent to the shortest distance between the points, which, as we all know, is a straight line. Thus, Fermat's principle is consistent with light traveling in a straight line in a homogeneous medium.

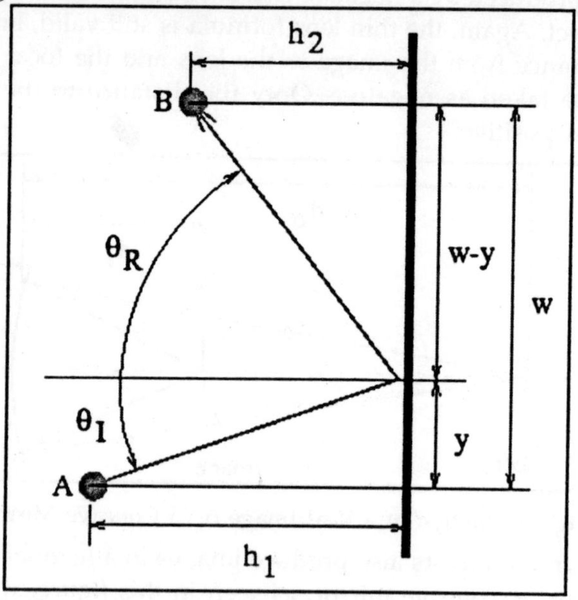

Fig. Definition sketch for deriving the law of reflection from Fermat's principle. θ_I is the angle of incidence and θ_R the angle of reflection.

Fermat's principle can also be used to derive the laws of reflection and refraction. For instance, figure shows a candidate ray for reflection in which the angles of incidence and reflection are not equal. The time required for the light to go from point A to point B is

$$t = \left[\left\{h_1^2 + y^2\right\}^{1/2} + \left\{h_2^2 + \left(w - y^2\right)\right\}^{1/2}\right]/c$$

where c is the speed of light. We find the minimum time by differentiating t with respect to y and setting the result to zero, with the result that

$$\frac{y}{\left\{h_2^2 + y^2\right\}^{1/2}} = \frac{\omega - y}{\left\{h_2^2 + \left(w - y^2\right)\right\}^{1/2}}.$$

However, we note that the left side of this equation is simply $\sin\theta_I$, while the right side is $\sin\theta_R$, so that the minimum time condition reduces to $\sin\theta_I = \sin\theta_R$ or $\theta_I = \theta_R$, which is the law of reflection.

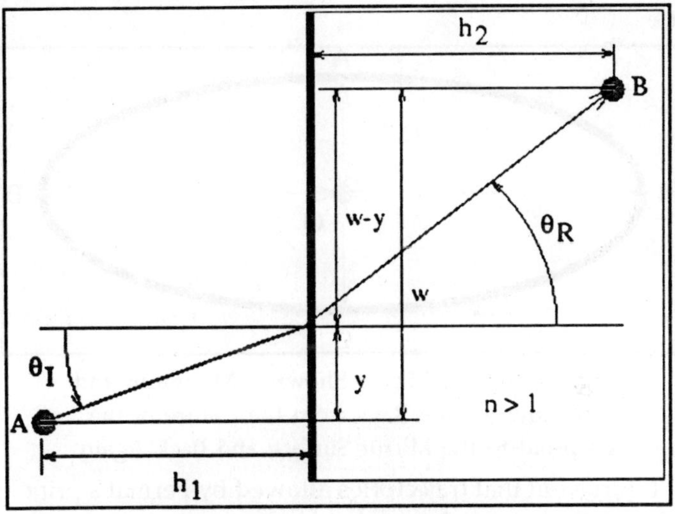

Fig. Definition sketch for deriving Snell's law of refraction from Fermat's principle. The shaded area has index of refraction $n > 1$.

A similar analysis may be done to derive Snell's law of refraction. The speed of light in a medium with refractive index n is c/n, where c is its speed in a vacuum. Thus, the time required for light to go some distance in such a medium is times the time light takes to go the same distance in a vacuum. Referring to figure, the time required for light to go from A to B becomes

$$t=\left[\left\{h_1^2+y^2\right\}^{1/2}+\left\{h_2^2+\left(w-y^2\right)\right\}^{1/2}\right]/c.$$

This results in the condition

$$\sin\theta_I = n\sin\theta_R$$

where θ_R is now the refracted angle. We recognize this result as Snell's law.

Notice that the reflection case illustrates a point about Fermat's principle: The minimum time may actually be a *local* rather than a global minimum – after all, the global minimum distance from A to B is still just a straight line between the two points! In fact, light starting from point A will reach point B by *both* routes – the direct route and the reflected route.

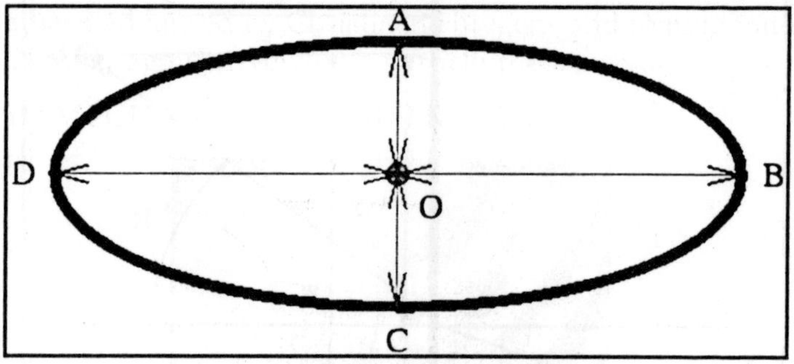

Fig. Ellipsoidal Mirror Showing Minimum and Maximum Time Rays from the Centre of the Ellipsoid to the Mirror Surface and back Again.

It turns out that trajectories allowed by Fermat's principle don't strictly have to be minimum time trajectories.

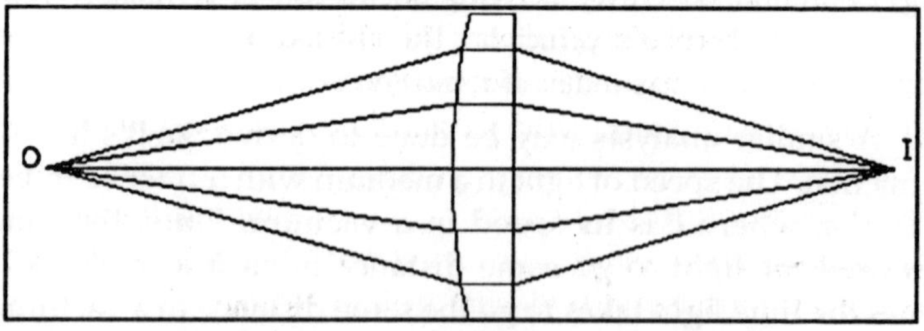

Fig. Ray Trajectories from a Point O Being Focused to Another Point I by a Lens.

They can also be *maximum* time trajectories. In this case light emitted at point O can be reflected back to point O from four points on the mirror, A, B, C, and D. The trajectories O-A-O and O-C-O are minimum time trajectories while O-B-O and O-D-O are maximum time trajectories.

Figure illustrates a rather peculiar situation. Notice that *all* the rays from point O which intercept the lens end up at point I. This would seem to contradict Fermat's principle, in that only the minimum (or maximum) time trajectories should occur.

However, a calculation shows that all the illustrated trajectories in this particular case take the same time. Thus, the light cannot choose one trajectory over another using Fermat's principle and all of the trajectories are equally favored. Note that this inference applies not to just any set of trajectories, but only those going from one focal point to another.

MATTER WAVES

We begin our study of quantum mechanics by discussing the diffraction undergone by X-rays and electrons when they interact with a crystal. X-rays are a form of electromagnetic radiation with wavelengths comparable to the distances between atoms.

Scattering from atoms in a regular crystalline structure results in an interference pattern which is in many ways similar to the pattern from a diffraction grating. We first develop Bragg's law for diffraction of X-rays from a crystal. Two practical techniques for doing X-ray diffraction are then described.

It turns out that electrons have wave-like properties and also undergo Bragg diffraction by crystals. Bragg diffraction thus provides a crucial bridge between the worlds of waves and particles. With this bridge we introduce the classical ideas of momentum and energy by relating them to the wave vector and frequency of a wave. The properties of waves also give rise to the Heisenberg uncertainty principle.

Table shows a table of the Nobel prizes associated with

the ideas presented. This gives us a feel for the chronology of these discoveries and indicates how important they were to the development of physics in the early 20th century.

Table. Selected Nobel Prize Winners, Year of Award, and Contribution.

Year	*Recipient*	*Contribution*
1901	W. K. Rφntgen	Discovery of X-rays
1906	J. J. Thomson	Discovery of electron
1914	M. von Laue	X-ray diffraction in crystals
1915	W. and L. Bragg	X-ray analysis of crystal structure
1918	M. Planck	Energy quantization
1921	A. Einstein	Photoelectric effect
1922	N. Bohr	Structure of atoms
1929	L.-V. de Broglie	Wave nature of electrons
1932	W. Heisenberg	Quantum mechanics
1933	Schrφdinger and Dirac	Atomic theory
1937	Davisson and Thomson	Electron diffraction in crystals

BRAGG'S LAW

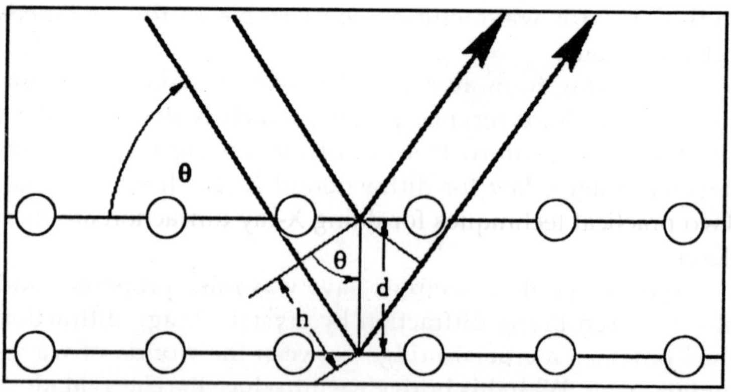

Fig. Schematic Diagram for Determining Bragg's Law.

Figure schematically illustrates interference between waves scattering from two adjacent rows of atoms in a crystal. The net effect of scattering from a single row is equivalent to partial reflection from a mirror imagined to be aligned with the row. Thus, the angle of "reflection" equals the angle of incidence for each row. Interference then occurs between the beams reflecting off different rows of atoms in the crystal.

For the two adjacent rows shown in figure, the path difference between beams is $2h = 2d\sin\theta$. For constructive interference this must be an integer number of wavelengths, $m\lambda$, where the integer is called the order of interference. The result is Bragg's law of diffraction:

$$m\lambda = 2d\sin\theta, \quad m = 1,2,3\ldots\text{(Bragg's law)}.$$

If only two rows are involved, the transition from constructive to destructive interference as θ changes is gradual. However, if interference from many rows occurs, then the constructive interference peaks become very sharp with mostly destructive interference in between. This sharpening of the peaks as the number of rows increases is very similar to the sharpening of the diffraction peaks from a diffraction grating as the number of slits increases.

X-RAY DIFFRACTION TECHNIQUES

Two types of targets are used in Bragg diffraction experiments, single crystals and powder targets.

Single Crystal

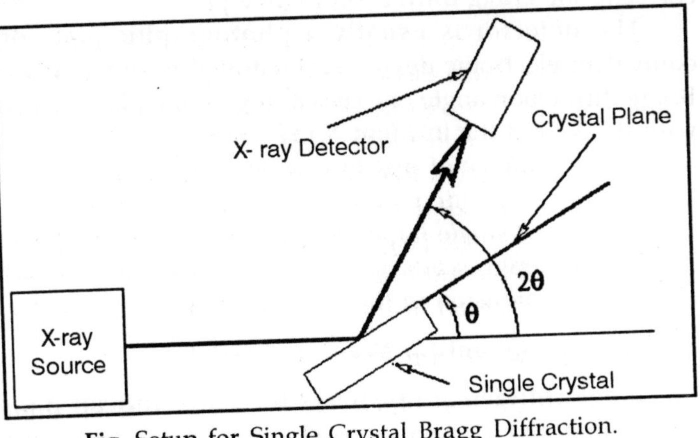

Fig. Setup for Single Crystal Bragg Diffraction.

In a single crystal setup, an X-ray detector is mounted as shown in figure. A mechanical device keeps the detector oriented so that the angle of incidence equals the angle of reflection for the desired crystal plane. Peaks in the X-ray detection rate are sought as the angle θ is varied.

The advantage of this type of apparatus is that diffraction peaks from only the selected crystal plane are observed.

Powder Target

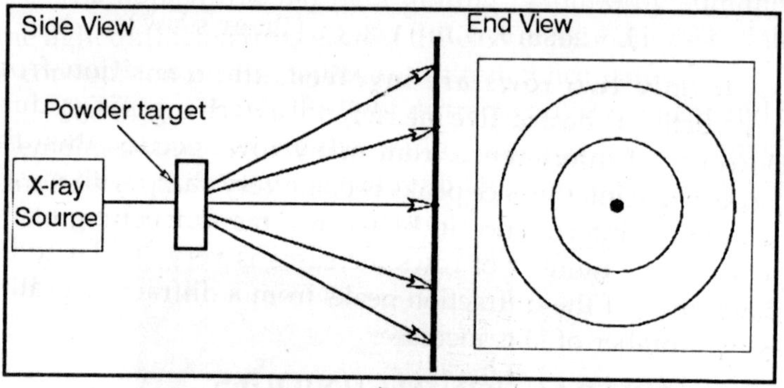

Fig. Setup for Powder Target Bragg Diffraction.

The powder in a powder target is really a conglomeration of many tiny crystals randomly oriented. Thus, for each possible Bragg diffraction angle there are crystals oriented correctly for Bragg diffraction to take place.

The detector is usually a photographic plate or an equivalent electronic device as illustrated in figure. For each Bragg diffraction angle one sees a ring on the plate concentric with the axis of the incident X-ray beam.

The advantage of this type of system is that no a priori knowledge is needed of the crystal plane orientations. Furthermore, a single large crystal is not required. However, all possible Bragg scattering angles are seen at once, which can lead to confusion in the interpretation of the results.

Meaning of Quantum Wave Function

Bragg diffraction illustrates the most difficult thing to understand about quantum mechanics, namely that particles can have wave-like properties and waves can have particle-like properties.

The variation of X-ray intensity with angle seen in a Bragg diffraction apparatus is very difficult to explain in any terms other than wave interference. Yet, X-rays are typically detected by a device such as a Geiger counter which produces a pulse of electricity for each X-ray particle, or *photon*, which hits it. Thus, X-rays sometimes act like particles and sometimes like waves.

Light isn't alone in having both particle and wave properties. Davisson and Germer and later G. P. Thomson (son of J. J. Thomson, the discoverer of the electron) showed that electrons also can act like waves. They did this by demonstrating that electrons undergo Bragg diffraction in crystals, much as X-rays do.

Most physicists (including Albert Einstein) have found quantum mechanics to be extremely bizarre, so if we feel the same way, we are in good company! However, there is a useful interpretation of quantum mechanics which at least allows us to get on with using it to solve problems, even though it may not satisfy our intuitive reservations about the theory.

The displacement of a quantum mechanical wave associated with a particle is usually called the wave function, ψ. Like light, it is not at all clear what ψ is a displacement *of*, but its use is straightforward. The absolute square of the wave function, $|\psi(x,t)|^2$, is proportional to the *probability* of finding the associated particle at position x and time t. The *absolute* square is taken because under many circumstances the wave function is actually *complex*, i. e., it has both real and imaginary parts. The reasons for this will be discussed later.

Due to the interpretation of the wave function, quantum mechanics is a probabilistic theory. It does not tell us with certainty what happens to a particular particle. Instead, it tells us, for instance, the probabilities for the particle to be detected in certain locations. If many experiments are done, with one particle per experiment, the numbers of experiments with particles being detected in the various possible locations are in proportion to the quantum mechanical probabilities.

Sense and Nonsense in Quantum Mechanics

The essential mystery of quantum mechanics becomes clearer when discussed in the conceptually simpler context of two slit interference.

If light and electrons can have both particle and wave properties, then one might ask through which of the two slits the particle passed. However, in physics a question simply doesn't make sense if it cannot be answered by experiment.

One can indeed perform an experiment to determine which slit the X-ray photon or electron passes through in the two slit experiment. However, by the very act of making this measurement, the form of the associated wave is altered. In particular, since the absolute square of the wave displacement represents the probability of finding the particle, once the particle has definitely been found passing through one or the other of the slits, the wave function collapses into a very small wave packet located at the observed position of the particle.

Thus, the wave displacement becomes zero at the slit it didn't pass through. However, the interference pattern results from the superposition of waves emanating from two slits. If no wave comes from one of the slits (because the wave displacement is zero there), then there can be no interference pattern!

We can now make the inverse argument. If there is an interference pattern, then we know that the wave displacement is non-zero at both slits.

From the probability interpretation of the wave displacement, we conclude that we cannot say, even in principle, through which slit the particle passed. It is not just that we don't know the answer to this question; there is simply no experiment which can give us an answer without destroying the interference pattern. In other words, the question "Which slit did the particle pass through?" is a nonsensical question in the case where an interference pattern is actually produced.

The American physicist Richard Feynman noticed that the above behaviour can be interpreted as violating the normal laws of probability. These laws say that the probability of an event is the sum of the probabilities of alternate independent ways for that event to occur.

For instance, the probability for a particle to reach point A on the detection screen of a two slit setup is just the probability P_1 for the particle to reach point A after going through slit 1, plus the probability P_2 for the particle to reach point A after going through slit 2.

Thus, if $P_1 = P_2 = 0.1$, then the probability for the particle to reach point A irrespective of which slit it went through should be $P_{total} = P_1 + P_2 = 0.2$. However, if point A happens to be a point of destructive interference, then we know that $P_{total} = 0$.

Feynman proposed that the above rule stating that alternate independent probabilities add, is simply incorrect. In its place Feynman asserted that probability amplitudes add instead, where the probability amplitude in this case is just the wave function associated with the particle.

The probability is obtained by adding the alternate probability amplitudes together and taking the absolute square of the sum.

In the two slit case, let us call ψ_1 the amplitude for the particle to reach point A via slit 1, while ψ_2 is the amplitude for it to reach this point via slit 2. The total amplitude is therefore $\psi_{total} = \psi_1 + \psi_2$ and the overall probability for the particle to reach point A is $P_{total} = |\psi_1 + \psi_2|^2$.

The wave functions ψ_1 and ψ_2 at point A are proportional respectively to $\sin(kL_1)$ and $\sin(kL_2)$, where k is the wavenumber and and L_1 are L_2 the respective distances from slit 1 and slit 2 to point A. It is easy to see that destructive interference occurs if the difference between L_1 and L_2 is half a wavelength, thus yielding $P_{total} = 0$.

The above treatment of two slits in terms of probability amplitudes yields results which are identical to the conventional treatment of the two slit problem for waves. However, it is conceptually clearer in the case of quantum mechanics because we no longer need to ask whether "things" are really particles or waves.

In particular, the particle/wave is replaced by just a particle plus the assertion that probability amplitudes rather than probabilities add.

The laws of quantum mechanics then boil down to rules for how probability amplitudes behave. In the case of particles, the amplitudes behave like waves, which explains why the study of wave kinematics and dynamics is so important to our understanding of quantum mechanics.

Chapter 12

Laser Physics

A laser is a device that emits light (electromagnetic radiation) through a process called *stimulated emission*. The term *laser* is an acronym for *light amplification by stimulated emission of radiation*. Laser light is usually spatially coherent, which means that the light either is emitted in a narrow, low-divergence beam, or can be converted into one with the help of optical components such as lenses.

Typically, lasers are thought of as emitting light with a narrow wavelength spectrum ("monochromatic" light). This is not true of all lasers, however: some emit light with a broad spectrum, while others emit light at multiple distinct wavelengths simultaneously. The coherence of typical laser emission is distinctive. Most other light sources emit *incoherent* light, which has a phase that varies randomly with time and position.

Terminology

The word *laser* originated as an acronym for *light amplification by stimulated emission of radiation*. The word *light* in this phrase is used in the broader sense, referring to electromagnetic radiation of any frequency, not just that in the visible spectrum. Hence there are *infrared lasers, ultraviolet lasers, X-ray lasers*, etc.

Because the microwave equivalent of the laser, the *maser*, was developed first, devices that emit microwave and radio frequencies are usually called *masers*. In early literature, particularly from researchers at Bell Telephone Laboratories, the laser was often called the *optical maser*. This usage has since become uncommon, and as of 1998 even Bell Labs uses the term *laser*.

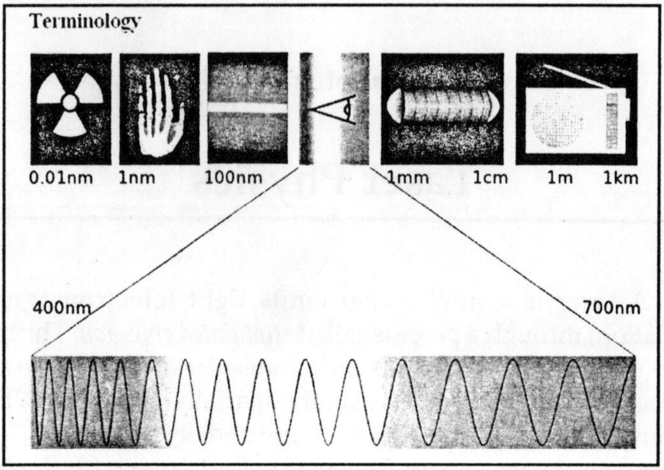

Fig. From Left to Right: Gamma Rays, X-rays, Ultraviolet Rays, Visible Spectrum, Infrared, Microwaves, Radio Waves.

The back-formed verb *to lase* means "to produce laser light" or "to apply laser light to". The word "laser" is sometimes used to describe other non-light technologies. For example, a source of atoms in a coherent state is called an "atom laser".

DESIGN

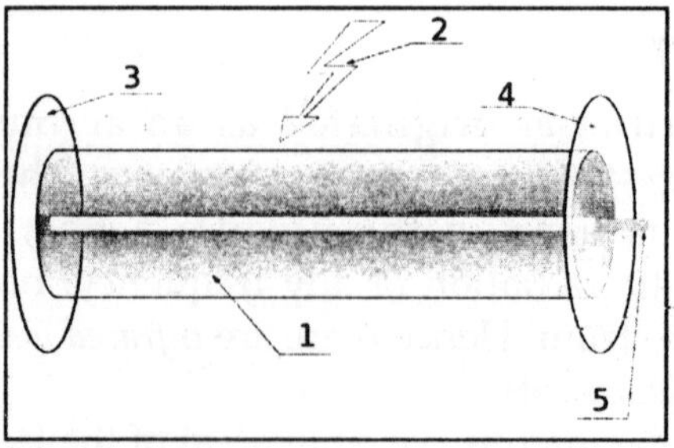

Principal Components:

1. Gain medium
2. Laser pumping energy
3. High reflector

4. Output coupler
5. Laser beam

A laser consists of a gain medium inside a highly reflective optical cavity, as well as a means to supply energy to the gain medium. The gain medium is a material with properties that allow it to amplify light by stimulated emission.

In its simplest form, a cavity consists of two mirrors arranged such that light bounces back and forth, each time passing through the gain medium. Typically one of the two mirrors, the output coupler, is partially transparent. The output laser beam is emitted through this mirror.

Light of a specific wavelength that passes through the gain medium is amplified (increases in power); the surrounding mirrors ensure that most of the light makes many passes through the gain medium, being amplified repeatedly. Part of the light that is between the mirrors (that is, within the cavity) passes through the partially transparent mirror and escapes as a beam of light.

The process of supplying the energy required for the amplification is called pumping. The energy is typically supplied as an electrical current or as light at a different wavelength. Such light may be provided by a flash lamp or perhaps another laser. Most practical lasers contain additional elements that affect properties such as the wavelength of the emitted light and the shape of the beam.

LASER PHYSICS

The glowing ray in the middle is an electric discharge producing light in much the same way as a neon light. It is the gain medium through which the laser passes, *not* the laser beam itself, which is visible there. The laser beam crosses the air and marks a red point on the screen to the right.

The gain medium of a laser is a material of controlled purity, size, concentration, and shape, which amplifies the beam by the process of stimulated emission. It can be of any state: gas, liquid, solid or plasma. The gain medium absorbs pump energy, which raises some electrons into higher-energy ("excited") quantum states.

Fig. A Helium-neon Laser

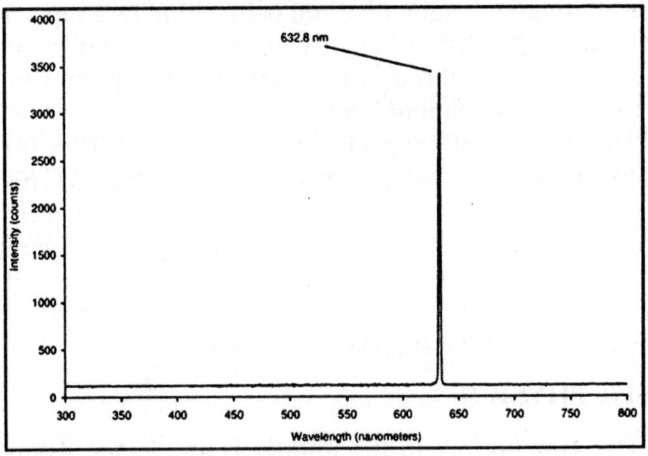

Fig. Spectrum of a Helium Neon Laser Showing the Very High Spectral Purity Intrinsic to Nearly all Lasers. Compare with the Relatively Broad Spectral Emittance of a Light Emitting Diode.

Particles can interact with light both by absorbing photons or by emitting photons. Emission can be spontaneous or stimulated. In the latter case, the photon is emitted in the same direction as the light that is passing by. When the number of particles in one excited state exceeds the number of particles in some lower-energy state, population inversion is achieved and the amount of stimulated emission due to light that passes through is larger than the amount of absorption. Hence, the light is amplified. By itself, this makes an optical amplifier. When an optical amplifier is placed inside a resonant optical cavity, one obtains a laser.

The light generated by stimulated emission is very similar to the input signal in terms of wavelength, phase, and polarization. This gives laser light its characteristic coherence, and allows it to maintain the uniform polarization and often monochromaticity established by the optical cavity design.

The optical cavity, a type of cavity resonator, contains a coherent beam of light between reflective surfaces so that the light passes through the gain medium more than once before it is emitted from the output aperture or lost to diffraction or absorption.

As light circulates through the cavity, passing through the gain medium, if the gain (amplification) in the medium is stronger than the resonator losses, the power of the circulating light can rise exponentially. But each stimulated emission event returns a particle from its excited state to the ground state, reducing the capacity of the gain medium for further amplification. When this effect becomes strong, the gain is said to be *saturated*.

The balance of pump power against gain saturation and cavity losses produces an equilibrium value of the laser power inside the cavity; this equilibrium determines the operating point of the laser. If the chosen pump power is too small, the gain is not sufficient to overcome the resonator losses, and the laser will emit only very small light powers.

The minimum pump power needed to begin laser action is called the *lasing threshold*. The gain medium will amplify any photons passing through it, regardless of direction; but only the photons aligned with the cavity manage to pass more than once through the medium and so have significant amplification.

The beam in the cavity and the output beam of the laser, if they occur in free space rather than waveguides (as in an optical fibre laser), are, at best, low order Gaussian beams.

However this is rarely the case with powerful lasers. If the beam is not a low-order Gaussian shape, the transverse modes of the beam can be described as a superposition of Hermite-Gaussian or Laguerre-Gaussian beams (for stable-cavity lasers).

Unstable laser resonators on the other hand, have been shown to produce fractal shaped beams. The beam may be highly *collimated*, that is being parallel without diverging.

However, a perfectly collimated beam cannot be created, due to diffraction. The beam remains collimated over a distance which varies with the square of the beam diameter, and eventually diverges at an angle which varies inversely with the beam diameter. Thus, a beam generated by a small laboratory laser such as a helium-neon laser spreads to about 1.6 kilometers (1 mile) diameter if shone from the Earth to the Moon.

By comparison, the output of a typical semiconductor laser, due to its small diameter, diverges almost as soon as it leaves the aperture, at an angle of anything up to 50°. However, such a divergent beam can be transformed into a collimated beam by means of a lens. In contrast, the light from non-laser light sources cannot be collimated by optics as well.

Although the laser phenomenon was discovered with the help of quantum physics, it is not essentially more quantum mechanical than other light sources. The operation of a free electron laser can be explained without reference to quantum mechanics.

MODES OF OPERATION

The output of a laser may be a continuous constant-amplitude output (known as *CW* or *continuous wave*); or pulsed, by using the techniques of Q-switching, modelocking, or gain-switching. In pulsed operation, much higher peak powers can be achieved.

Some types of lasers, such as *dye lasers* and *vibronic solid-state lasers* can produce light over a broad range of wavelengths; this property makes them suitable for generating extremely short pulses of light, on the order of a few femtoseconds (10^{-15} s).

Continuous Wave Operation

In the continuous wave (CW) mode of operation, the output of a laser is relatively constant with respect to time. The population inversion required for lasing is continually maintained by a steady pump source.

Pulsed Operation

In the pulsed mode of operation, the output of a laser varies with respect to time, typically taking the form of alternating 'on' and 'off' periods. In many applications one aims to deposit as much energy as possible at a given place in as short time as possible.

In laser ablation for example, a small volume of material at the surface of a work piece might evaporate if it gets the energy required to heat it up far enough in very short time. If, however, the same energy is spread over a longer time, the heat may have time to disperse into the bulk of the piece, and less material evaporates. There are a number of methods to achieve this.

Q-SWITCHING

In a Q-switched laser, the population inversion (usually produced in the same way as CW operation) is allowed to build up by making the cavity conditions (the 'Q') unfavorable for lasing. Then, when the pump energy stored in the laser medium is at the desired level, the 'Q' is adjusted (electro- or acousto-optically) to favourable conditions, releasing the pulse. This results in high peak powers as the average power of the laser (were it running in CW mode) is packed into a shorter time frame.

Q-switching, sometimes known as giant pulse formation, is a technique by which a laser can be made to produce a pulsed output beam. The technique allows the production of light pulses with extremely high (gigawatt) peak power, much higher than would be produced by the same laser if it were operating in a continuous wave (constant output) mode. Compared to modelocking, another technique for pulse generation with lasers, Q-switching leads to much lower pulse repetition rates, much higher pulse energies, and much longer pulse durations. Both techniques are sometimes applied at once.

Q-switching was first proposed in 1958 by Gordon Gould, and independently discovered and demonstrated in 1961 or 1962 by R.W. Hellwarth and F.J. McClung using electrically switched Kerr cell shutters in a ruby laser.

PRINCIPLE OF Q-SWITCHING

Q-switching is achieved by putting some type of variable attenuator inside the laser's optical resonator. When the attenuator is functioning, light which leaves the gain medium does not return, and lasing cannot begin. This attenuation inside the cavity corresponds to a decrease in the *Q factor* or *quality factor* of the optical resonator. A high Q factor corresponds to low resonator losses per roundtrip, and vice versa. The variable attenuator is commonly called a "Q-switch", when used for this purpose.

Initially the laser medium is pumped while the Q-switch is set to prevent feedback of light into the gain medium (producing an optical resonator with low Q). This produces a population inversion, but laser operation cannot yet occur since there is no feedback from the resonator. Since the rate of stimulated emission is dependent on the amount of light entering the medium, the amount of energy stored in the gain medium increases as the medium is pumped. Due to losses from spontaneous emission and other processes, after a certain time the stored energy will reach some maximum level; the medium is said to be *gain saturated.*

At this point, the Q-switch device is quickly changed from low to high Q, allowing feedback and the process of optical amplification by stimulated emission to begin. Because of the large amount of energy already stored in the gain medium, the intensity of light in the laser resonator builds up very quickly; this also causes the energy stored in the medium to be depleted almost as quickly. The net result is a short pulse of light output from the laser, known as a *giant pulse,* which may have a very high peak intensity.

There are two main types of Q-switching:

Active Q-Switching

Here, the Q-switch is an externally-controlled variable attenuator. This may be a mechanical device such as a shutter, chopper wheel or spinning mirror placed inside the cavity, or (more commonly) it may be some form of modulator such as an acousto-optic device or an electro-optic device – a Pockels cell or Kerr cell. The reduction of losses (increase of Q) is triggered by an external event, typically an electrical signal. The pulse repetition rate can therefore be externally controlled.

Modulators generally allow a faster transition from low to high Q, and provide better control. An additional advantage of modulators is that the rejected light may be coupled out of the cavity and can be used for something else. Alternatively, when the modulator is in its low-Q state, an externally-generated beam can be coupled *into* the cavity through the modulator. This can be used to "seed" the cavity with a beam that has desired characteristics (such as transverse mode or wavelength). When the Q is raised, lasing builds up from the initial seed, producing a Q-switched pulse that has characteristics inherited from the seed.

Passive Q-Switching

In this case, the Q-switch is a saturable absorber, a material whose transmission increases when the intensity of light exceeds some threshold. The material may be an ion-doped crystal like Cr:YAG, which is used for Q-switching of Nd:YAG lasers, a bleachable dye, or a passive semiconductor device. Initially, the loss of the absorber is high, but still low enough to permit some lasing once a large amount of energy is stored in the gain medium.

As the laser power increases, it saturates the absorber, i.e., rapidly reduces the resonator loss, so that the power can increase even faster. Ideally, this brings the absorber into a state with low losses to allow efficient extraction of the stored energy by the laser pulse.

After the pulse, the absorber recovers to its high-loss state before the gain recovers, so that the next pulse is delayed until the energy in the gain medium is fully replenished. The pulse repetition rate can only indirectly be controlled, e.g. by varying the laser's pump power and the amount of saturable absorber in the cavity. Direct control of the repetition rate can be achieved by using a pulsed pump source as well as passive Q-switching.

Variants

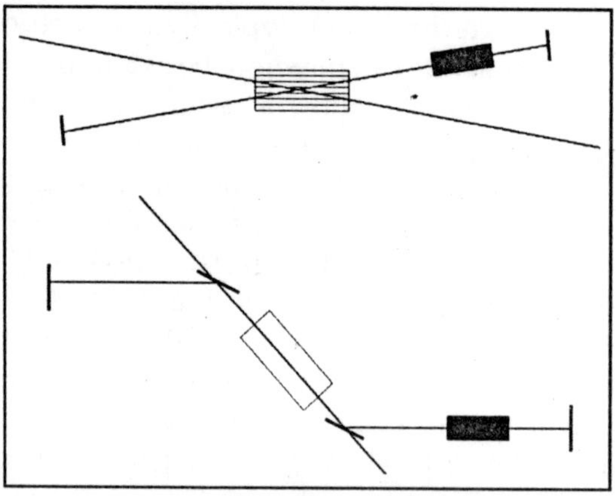

Fig. Regenerative Amplifier.

The Pockel's cell based design needs thin film polarizers. The direction of the emitted pulse depends on the timing.

- Jitter can be reduced by not reducing the Q by as much, so that a small amount of light can still circulate in the cavity. This provides a "seed" of light that can aid in the buildup of the next Q-switched pulse.
- Cavity dumping: The cavity end mirrors are 100% reflective, so that no output beam is produced when the Q is high. Instead, the Q-switch is used to "dump" the beam out of the cavity after a time delay. The cavity Q goes from low to high to start the

laser buildup, and then goes from high to low to "dump" the beam from the cavity all at once. This produces a shorter output pulse than regular Q-switching. Electro-optic modulators are normally used for this, since they can easily be made to function as a near-perfect beam "switch" to couple the beam out of the cavity. The modulator that dumps the beam may be the same modulator that Q-switches the cavity, or a second (possibly identical) modulator. A dumped cavity is more complicated to align than simple Q-switching, and may need a control loop to choose the best time at which to dump the beam from the cavity.

- Regenerative amplification: In regenerative amplification, an optical amplifier is placed inside a Q-switched cavity. Pulses of light from another laser (the "master oscillator") are injected into the cavity by lowering the Q to allow the pulse to enter and then increasing the Q to confine the pulse to the cavity where it can be amplified by repeated passes through the gain medium. The pulse is then allowed to leave the cavity via another Q switch.

Typical Performance

A typical Q-switched laser (e.g. a Nd:YAG laser) with a resonator length of e.g. 10 cm can produce light pulses of several tens of nanoseconds duration. Even when the average power is well below 1 W, the peak power can be many kilowatts. Large-scale laser systems can produce Q-switched pulses with energies of many joules and peak powers in the gigawatt region. On the other hand, passively Q-switched microchip lasers (with very short resonators) have generated pulses with durations far below one nanosecond and pulse repetition rates from hundreds of hertz to several megahertz (MHz)

Applications

Q-switched lasers are often used in applications which

demand high laser intensities in nanosecond pulses, such metal cutting or pulsed holography. Nonlinear optics often takes advantage of the high peak powers of these lasers, offering applications such as 3D optical data storage and 3D microfabrication. However, Q-switched lasers can also be used for measurement purposes, such as for distance measurements (range finding) by measuring the time it takes for the pulse to get to some target and the reflected light to get back to the sender.

Q-switched lasers are also used to remove tattoos. They are used to shatter tattoo pigment into particles that are cleared by the body's lymphatic system. Full removal takes an average of eight treatments, spaced at least a month apart, using different wavelengths for different colored inks.

MODELOCKING

A modelocked laser emits extremely short pulses on the order of tens of picoseconds down to less than 10 femtoseconds. These pulses are typically separated by the time that a pulse takes to complete one round trip in the resonator cavity.

Due to the Fourier limit (also known as energy-time uncertainty), a pulse of such short temporal length has a spectrum which contains a wide range of wavelengths. Because of this, the laser medium must have a broad enough gain profile to amplify them all. An example of a suitable material is titanium-doped, artificially grown sapphire (Ti:sapphire).

The modelocked laser is a most versatile tool for researching processes happening at extremely fast time scales also known as femtosecond physics, femtosecond chemistry and ultrafast science, for maximizing the effect of nonlinearity in optical materials (e.g. in second-harmonic generation, parametric down-conversion, optical parametric oscillators and the like), and in ablation applications. Again, because of the short timescales involved, these lasers can achieve extremely high powers.

Mode-locking is a technique in optics by which a laser can be made to produce pulses of light of extremely short duration, on the order of picoseconds (10^{-12}s) or femtoseconds (10^{-15}s).

The basis of the technique is to induce a fixed phase relationship between the modes of the laser's resonant cavity. The laser is then said to be *phase-locked* or *mode-locked*. Interference between these modes causes the laser light to be produced as a train of pulses. Depending on the properties of the laser, these pulses may be of extremely brief duration, as short as a few femtoseconds.

LASER CAVITY MODES

Although laser light is perhaps the purest form of light, it is not of a single, pure frequency or wavelength. All lasers produce light over some natural bandwidth or range of frequencies. A laser's bandwidth of operation is determined primarily by the gain medium that the laser is constructed from, and the range of frequencies that a laser may operate over is known as the *gain bandwidth*.

For example, a typical helium-neon (HeNe) gas laser has a gain bandwidth of approximately 1.5 GHz (a wavelength range of about 0.002 nm at a central wavelength of 633 nm), whereas a titanium-doped sapphire (Ti:Sapphire) solid-state laser has a bandwidth of about 128 THz (a 300 nm wavelength range centred around 800 nm).

The second factor that determines a laser's emission frequencies is the optical cavity or resonant cavity of the laser. In the simplest case, this consists of two plane (flat) mirrors facing each other, surrounding the gain medium of the laser (this arrangement is known as a Fabry-Perot cavity). Since light is a wave, when bouncing between the mirrors of the cavity the light will constructively and destructively interfere with itself, leading to the formation of standing waves between the mirrors.

These standing waves form a discrete set of frequencies, known as the *longitudinal modes* of the cavity. These modes are the only frequencies of light which are self-regenerating and allowed to oscillate by the resonant cavity; all other frequencies of light are suppressed by destructive interference. For a simple plane-mirror cavity, the allowed modes are those for which the separation distance of the mirrors L is an exact multiple of half the wavelength of the light λ, such that $L = q\,\lambda/2$, when q is an integer known as the mode order.

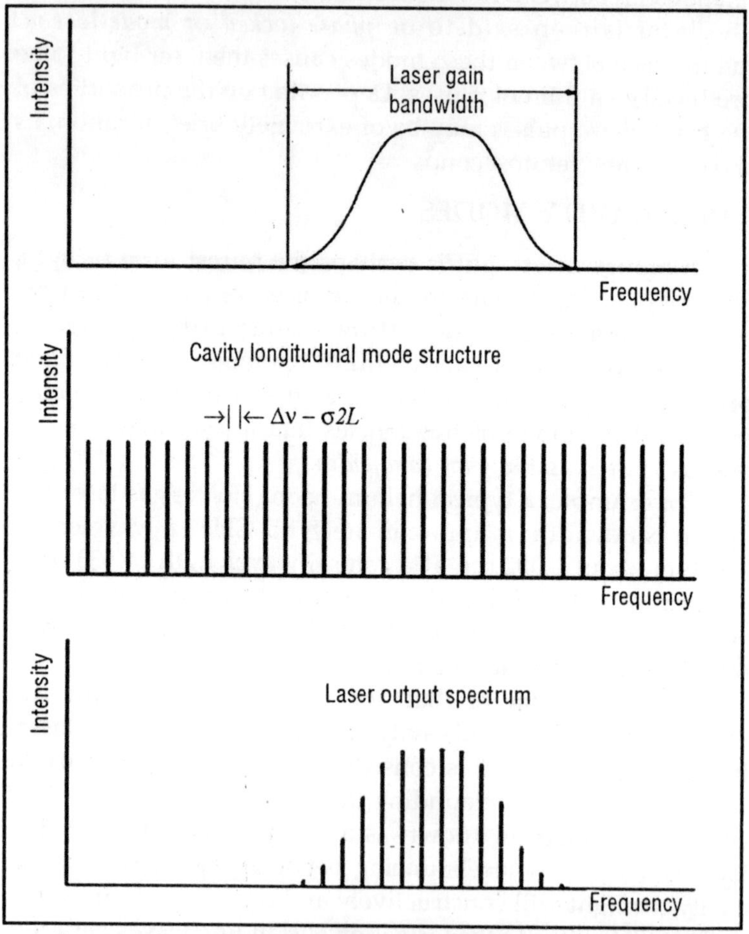

In practice, the separation distance of the mirrors L is usually much greater than the wavelength of light λ, so the relevant values of q are large (around 10^5 to 10^6). Of more interest is the frequency separation between any two adjacent modes q and $q+1$; this is given (for an empty linear resonator of length L) by $\Delta\nu$:

$$\Delta\nu = \frac{c}{2L}$$

where c is the speed of light ($H \approx 3\times10^8$ m $\cdot$s^{-1}).

Using the equation, a small laser with a mirror separation of 30 cm has a frequency separation between longitudinal modes of 0.5 GHz. Thus for the two lasers referenced, with a 30 cm cavity the 1.5 GHz bandwidth of the HeNe laser would support up to 3 longitudinal modes, whereas the 128 THz bandwidth of the Ti:sapphire laser could support approximately 250,000 modes.

When more than one longitudinal modes are excited, the laser is said to be in multi-mode operation. When only one longitudinal mode is excited, the laser is said to be in single-mode operation.

Each individual longitudinal mode has itself some bandwidth or narrow range of frequencies over which it operates, but typically this bandwidth, determined by the finesse of the cavity, is much smaller than the inter-mode frequency separation.

MODE-LOCKING THEORY

In a simple laser, each of these modes will oscillate independently, with no fixed relationship between each other, in essence like a set of independent lasers all emitting light at slightly different frequencies. The individual phase of the light waves in each mode is not fixed, and may vary randomly due to such things as thermal changes in materials of the laser.

In lasers with only a few oscillating modes, interference between the modes can cause beating effects in the laser output, leading to random fluctuations in intensity; in lasers with many thousands of modes, these interference effects tend to average to a near-constant output intensity, and the laser operation is known as a *c.w.* or *continuous wave*.

If instead of oscillating independently, each mode operates with a fixed phase between it and the other modes, the laser output behaves quite differently. Instead of a random or constant output intensity, the modes of the laser will periodically all constructively interfere with one another, producing an intense burst or pulse of light. Such a laser is said to be *mode-locked* or *phase-locked*. These pulses occur separated in time by $\tau = 2L/c$, where τ is the time taken for the light to make exactly one round trip of the laser cavity. This time corresponds to a frequency exactly equal to the mode spacing of the laser, $\Delta\nu = 1/\tau$.

The duration of each pulse of light is determined by the number of modes which are oscillating in phase (in a real laser, it is not necessarily true that *all* of the laser's modes will be phase-locked). If there are N modes locked with a frequency separation $\Delta\nu$, the overall modelocked bandwidth is $N\Delta\nu$, and the wider this bandwidth, the shorter the pulse duration from the laser. In practice, the actual pulse duration is determined by the shape of each pulse, which is in turn determined by the exact amplitude and phase relationship of each longitudinal mode. For example, for a laser producing pulses with a Gaussian temporal shape, the minimum possible pulse duration Δt is given by

$$\Delta t = \frac{0.44}{N\Delta\nu}.$$

The value 0.44 is known as the *time-bandwidth product* of the pulse, and varies depending on the pulse shape. For ultrashort pulse lasers, a hyperbolic-secant-squared (sech^2) pulse shape is often assumed, giving a time-bandwidth product of 0.315.

Using this equation, we can calculate the minimum pulse duration which can be produced by a laser. For the HeNe laser with a 1.5 GHz bandwidth, the shortest Gaussian pulse which can be produced would be around 300 picoseconds; for the 128 THz bandwidth Ti:sapphire laser, this duration would be only 3.4 femtoseconds. These values represent the shortest possible Gaussian pulses supported by the laser's bandwidth; in a real mode-locked laser, the actual pulse duration depends on many other factors, such as the actual pulse shape, and the overall dispersion of the cavity.

Mode-Locking Methods

Methods for producing mode-locking in a laser may be

classified as either *active* or *passive*. Active methods typically involve using an external signal to induce a modulation of the intra-cavity light. Passive methods do not use an external signal, but rely on placing some element into the laser cavity which causes self-modulation of the light.

Active Mode-Locking

The most common active mode-locking technique places a standing wave acousto-optic modulator into the laser cavity. When driven with an electrical signal, this produces a sinusoidal amplitude modulation of the light in the cavity.

Considering this in the frequency domain, if a mode has optical frequency ν, and is amplitude-modulated at a frequency f, the resulting signal has sidebands at optical frequencies ν-f and ν+f. If the modulator is driven at the same frequency as the cavity-mode spacing $\Delta\nu$, then these sidebands correspond to the two cavity modes adjacent to the original mode. Since the sidebands are driven in-phase, the central mode and the adjacent modes will be phase-locked together.

Further operation of the modulator on the sidebands produces phase-locking of the ν-2f and ν+2f modes, and so on until all modes in the gain bandwidth are locked. Typical lasers are multi-mode and not seeded by a root mode. So multiple modes need to work out which phase to use. In a passive cavity with this locking applied there is no way to dump the entropy given by the original independent phases. This locking is better described as a coupling, leading to a complicated behaviour and not clean pulses. The coupling is only dissipative because of the dissipative nature of the amplitude modulation. Otherwise the phase modulation would not work.

This process can also be considered in the time domain. The amplitude modulator acts as a weak *shutter* to the light bouncing between the mirrors of the cavity, attenuating the light when it is "closed", and letting it through when it is "open". If the modulation rate f is synchronised to the cavity round-trip time τ, then a single pulse of light will bounce back and forth in the cavity. The actual strength of the modulation does not have to be large; a modulator that attenuates 1% of the light when "closed" will mode-lock a laser, since the same part of the light is repeatedly attenuated as it traverses the cavity.

Related to this amplitude modulation (AM) active mode-locking is frequency modulation (FM) mode-locking, which uses a modulator device based on the electro-optic effect. This device, when placed in a laser cavity and driven with an electrical signal, induces a small, sinusoidally varying frequency shift in the light passing through it.

If the frequency of modulation is matched to the round-trip time of the cavity, then some light in the cavity sees repeated up-shifts in frequency, and some repeated down-shifts. After many repetitions, the up-shifted and down-shifted light is swept out of the gain bandwidth of the laser. The only light which is unaffected is that which passes through the modulator when the induced frequency shift is zero, which forms a narrow pulse of light.

The third method of active mode-locking is synchronous mode-locking, or synchronous pumping. In this, the pump source (energy source) for the laser is itself modulated, effectively turning the laser on and off to produce pulses. Typically, the pump source is itself another mode-locked laser. This technique requires accurately matching the cavity lengths of the pump laser and the driven laser.

PASSIVE MODE-LOCKING

Passive mode-locking techniques are those that do not require a signal external to the laser (such as the driving signal of a modulator) to produce pulses. Rather, they use the light in the cavity to cause a change in some intracavity element, which will then itself produce a change in the intracavity light. The most common type of device which will do this is a saturable absorber.

A saturable absorber is an optical device that exhibits an intensity-dependent transmission. What this means is that the device behaves differently depending on the intensity of the light passing through it. For passive mode-locking, ideally a saturable absorber will selectively absorb low-intensity light, and transmit light which is of sufficiently high intensity.

When placed in a laser cavity, a saturable absorber will attenuate low-intensity constant wave light (pulse wings). However, because of the somewhat random intensity fluctuations experienced by an un-modelocked laser, any random, intense spike will be transmitted preferentially by the saturable absorber. As the light in the cavity oscillates, this process repeats, leading to the selective amplification of the high-intensity spikes, and the absorption of the low-intensity light. After many round trips, this leads to a train of pulses and mode-locking of the laser.

Considering this in the frequency domain, if a mode has optical frequency ν, and is amplitude-modulated at a frequency $n f$, the resulting signal has sidebands at optical frequencies $\nu - n f$ and $\nu + n f$ and enables much stronger mode-locking for shorter pulses and more stability than active modelocking, but has startup problems.

Saturable absorbers are commonly liquid organic dyes, but they can also be made from doped crystals and semiconductors. Semiconductor absorbers tend to exhibit very fast response times (~100 fs), which is one of the factors that determines the final duration of the pulses in a passively mode-locked laser. In a *colliding-pulse mode-locked laser* the absorber steepens the leading edge while the lasing medium steepens the trailing edge of the pulse.

There are also passive mode-locking schemes that do not rely on materials that directly display an intensity dependent absorption. In these methods, nonlinear optical effects in intra-cavity components are used to provide a method of selectively amplifying high-intensity light in the cavity, and attenuation of low-intensity light.

One of the most successful schemes is called Kerr-lens mode-locking (KLM), also sometimes called "self mode-locking". This uses a nonlinear optical process, the optical Kerr effect, which results in high-intensity light being focussed differently than low-intensity light. By careful arrangement of an aperture in the laser cavity, this effect can be exploited to produce the equivalent of an ultra-fast response time saturable absorber.

PRACTICAL MODE-LOCKED LASERS

In practice, a number of design considerations affect the performance of a mode-locked laser. The most important are the overall dispersion of the laser's optical resonator, which can be controlled with a prism compressor or some dispersive mirrors placed in the cavity, and optical nonlinearities.

For excessive net group delay dispersion (GDD) of the laser cavity, the phase of the cavity modes can not be locked over a large bandwidth, and it will be difficult to obtain very short pulses.

For a suitable combination of negative (anomalous) net GDD with the Kerr nonlinearity, soliton-like interactions may stabilize the mode-locking and help to generate shorter pulses. The shortest possible pulse duration is usually accomplished either for zero dispersion (without nonlinearities) or for some slightly negative (anomalous) dispersion (exploiting the soliton mechanism).

The shortest directly produced optical pulses are generally produced by Kerr-lens mode-locked Ti-sapphire lasers, and are around 5 femtoseconds long. Alternatively, amplified pulses of a similar duration are created through the compression of longer (e.g. 30 fs) pulses via self-phase modulation in a hollow core fibre or during filamentation. However, the minimum pulse duration is limited by the period of the carrier frequency (which is about 2.7 fs for Ti:S systems), therefore shorter pulses require moving to shorter wavelengths.

Some advanced techniques (involving high harmonic generation with amplified femtosecond laser pulses) can be used to produce optical features with durations as short as 100 attoseconds in the extreme ultraviolet spectral region (i.e. <30 nm).

Other achievements, important particularly for laser applications, concern the development of mode-locked lasers which can be pumped with laser diodes, can generate very high average output powers (tens of watts) in sub-picosecond pulses, or generate pulse trains with extremely high repetition rates of many GHz.

Pulse durations less than approximately 100 fs are too short to be directly measured using optoelectronic techniques (i.e. photodiodes), and so indirect methods such as autocorrelation, Frequency-resolved optical gating, Spectral phase interferometry for direct electric-field reconstruction or Multi-photon Intrapulse Interference Phase Scan are used.

PULSED PUMPING

Another method of achieving pulsed laser operation is to pump the laser material with a source that is itself pulsed, either through electronic charging in the case of flashlamps, or another laser which is already pulsed. Pulsed pumping was historically used with dye lasers where the inverted population lifetime of a dye molecule was so short that a high energy, fast pump was needed.

The way to overcome this problem was to charge up large capacitors which are then switched to discharge through flashlamps, producing a broad spectrum pump flash. Pulsed pumping is also required for lasers which disrupt the gain medium so much during the laser process that lasing has to cease for a short period. These lasers, such as the excimer laser and the copper vapour laser, can never be operated in CW mode.

FOUNDATIONS

In 1917 Albert Einstein, in his paper Zur Quantentheorie der Strahlung (On the Quantum Theory of Radiation), laid the foundation for the invention of the laser and its predecessor, the maser, in a ground-breaking rederivation of Max Planck's law of radiation based on the concepts of probability coefficients (later to be termed 'Einstein coefficients') for the absorption, spontaneous emission, and stimulated emission of electromagnetic radiation.

In 1928 Rudolph W. Landenburg confirmed the existence of stimulated emission and negative absorption. In 1939, Valentin A. Fabrikant predicted the use of stimulated emission to amplify "short" waves.

In 1947 Willis E. Lamb and R. C. Retherford found apparent stimulated emission in hydrogen spectra and made the first demonstration of stimulated emission.

In 1950 Alfred Kastler proposed the method of optical pumping, which was experimentally confirmed by Brossel, Kastler and Winter two years later.

The first working laser was demonstrated on 16 May 1960 by Theodore Maiman at Hughes Research Laboratories. Since then, lasers have become a multi-billion dollar industry. By far the largest single application of lasers is in optical storage devices such as compact disc and DVD players, in which a semiconductor laser less than a millimeter wide scans the surface of the disc. The second-largest application is fibre-optic communication. Other common applications of lasers are bar code readers, laser printers and laser pointers.

Maser

In 1953, Charles H. Townes and graduate students James P. Gordon and Herbert J. Zeiger produced the first microwave amplifier, a device operating on similar principles to the laser, but amplifying microwave rather than infrared or visible radiation. Townes's maser was incapable of continuous output. Nikolay Basov and Aleksandr Prokhorov of the Soviet Union worked independently on the quantum oscillator and solved the problem of continuous output systems by using more than two energy levels and produced the first maser.

These systems could release stimulated emission without falling to the ground state, thus maintaining a population inversion. In 1955 Prokhorov and Basov suggested an optical pumping of multilevel system as a method for obtaining the population inversion, which later became one of the main methods of laser pumping.

Townes reports that he encountered opposition from a number of eminent colleagues who thought the maser was theoretically impossible – including Niels Bohr, John von Neumann, Isidor Rabi, Polykarp Kusch, and Llewellyn H. Thomas.

Townes, Basov, and Prokhorov shared the Nobel Prize in Physics in 1964 "For fundamental work in the field of quantum electronics, which has led to the construction of oscillators and amplifiers based on the maser-laser principle".

Laser

In 1957, Charles Hard Townes and Arthur Leonard Schawlow, then at Bell Labs, began a serious study of the infrared laser. As ideas were developed, infrared frequencies were abandoned with focus on visible light instead. The concept was originally known as an "optical maser". Bell Labs filed a patent application for their proposed optical maser a year later. Schawlow and Townes sent a manuscript of their theoretical calculations to Physical Review, which published their paper that year.

At the same time Gordon Gould, a graduate student at Columbia University, was working on a doctoral thesis on the energy levels of excited thallium. Gould and Townes met and had conversations on the general subject of radiation emission. Afterwards Gould made notes about his ideas for a "laser" in November 1957, including suggesting using an open resonator, which became an important ingredient of future lasers.

In 1958, Prokhorov independently proposed using an open resonator, the first published appearance of this idea. Schawlow and Townes also settled on an open resonator design, apparently unaware of both the published work of Prokhorov and the unpublished work of Gould.

The term "laser" was first introduced to the public in Gould's 1959 conference paper "The LASER, Light Amplification by Stimulated Emission of Radiation". Gould intended "-aser" to be a suffix, to be used with an appropriate prefix for the spectrum of light emitted by the device (x-rays:

xaser, ultraviolet: *uvaser*, etc.). None of the other terms became popular, although "raser" was used for a short time to describe radio-frequency emitting devices.

Gould's notes included possible applications for a laser, such as spectrometry, interferometry, radar, and nuclear fusion. He continued working on his idea and filed a patent application in April 1959. The U.S. Patent Office denied his application and awarded a patent to Bell Labs in 1960. This sparked a legal battle that ran 28 years, with scientific prestige and much money at stake. Gould won his first minor patent in 1977, but it was not until 1987 that he could claim his first significant patent victory when a Federal judge ordered the government to issue patents to him for the optically pumped laser and the gas discharge laser.

The first working laser was made by Theodore H. Maiman in 1960 at Hughes Research Laboratories in Malibu, California, beating several research teams including those of Townes at Columbia University, Arthur Schawlow at Bell Labs, and Gould at a company called TRG (Technical Research Group). Maiman used a solid-state flashlamp-pumped synthetic ruby crystal to produce red laser light at 694 nanometres wavelength. Maiman's laser, however, was only capable of pulsed operation due to its three-level pumping scheme.

Later in 1960 the Iranian physicist Ali Javan, working with William R. Bennett and Donald Herriot, made the first gas laser using helium and neon. Javan later received the Albert Einstein Award in 1993.

The concept of the semiconductor laser diode was proposed by Basov and Javan. The first *laser diode* was demonstrated by Robert N. Hall in 1962. Hall's device was made of gallium arsenide and emitted at 850 nm in the near-infrared region of the spectrum. The first semiconductor laser with visible emission was demonstrated later the same year by Nick Holonyak, Jr. As with the first gas lasers, these early semiconductor lasers could be used only in pulsed operation, and indeed only when cooled to liquid nitrogen temperatures (77 K).

In 1970, Zhores Alferov in the Soviet Union and Izuo Hayashi and Morton Panish of Bell Telephone Laboratories independently developed laser diodes continuously operating at room temperature, using the heterojunction structure.

RECENT INNOVATIONS

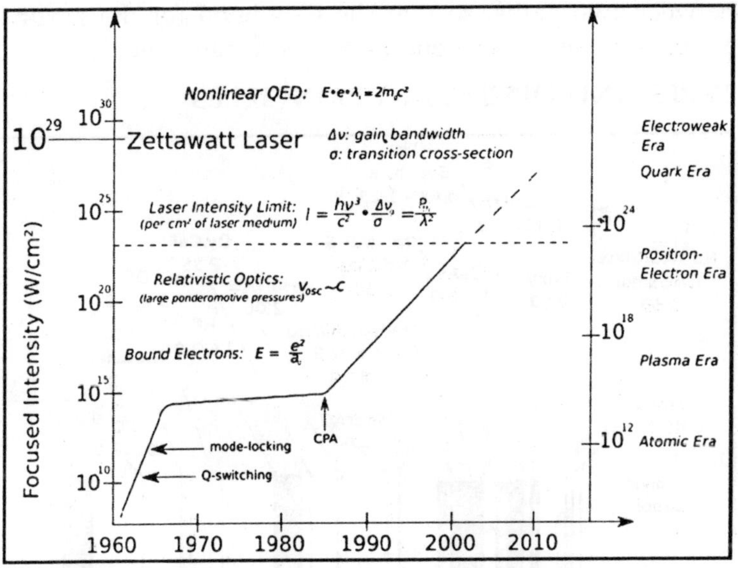

Fig. Graph Showing the History of Maximum Laser Pulse Intensity Throughout the Past 40 Years.

Since the early period of laser history, laser research has produced a variety of improved and specialized laser types, optimized for different performance goals, including:

- New wavelength bands
- Maximum average output power
- Maximum peak output power
- Minimum output pulse duration
- Maximum power efficiency
- Maximum charging
- Maximum firing
- Minimum cost

and this research continues to this day.

Lasing without maintaining the medium excited into a

population inversion was discovered in 1992 in sodium gas and again in 1995 in rubidium gas by various international teams.

This was accomplished by using an external maser to induce "optical transparency" in the medium by introducing and destructively interfering the ground electron transitions between two paths, so that the likelihood for the ground electrons to absorb any energy has been cancelled.

TYPES AND OPERATING PRINCIPLES

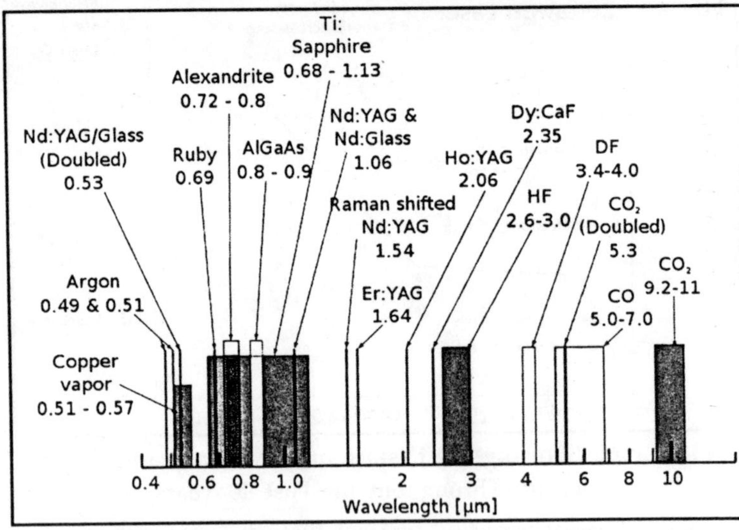

Fig. Spectral Output of Several Types of Lasers.

Gas Lasers

Gas lasers using many gases have been built and used for many purposes.

The helium-neon laser (HeNe) emits at a variety of wavelengths and units operating at 633 nm are very common in education because of its low cost.

Carbon dioxide lasers can emit hundreds of kilowatts at 9.6 µm and 10.6 µm, and are often used in industry for cutting and welding. The efficiency of a CO_2 laser is over 10%.

Argon-ion lasers emit light in the range 351-528.7 nm.

Depending on the optics and the laser tube a different number of lines is usable but the most commonly used lines are 458 nm, 488 nm and 514.5 nm.

A nitrogen transverse electrical discharge in gas at atmospheric pressure (TEA) laser is an inexpensive gas laser producing UV Light at 337.1 nm.

Metal ion lasers are gas lasers that generate deep ultraviolet wavelengths. Helium-silver (HeAg) 224 nm and neon-copper (NeCu) 248 nm are two examples. These lasers have particularly narrow oscillation linewidths of less than 3 GHz (0.5 picometers), making them candidates for use in fluorescence suppressed Raman spectroscopy.

Chemical Lasers

Chemical lasers are powered by a chemical reaction, and can achieve high powers in continuous operation. For example, in the Hydrogen fluoride laser (2700-2900 nm) and the Deuterium fluoride laser (3800 nm) the reaction is the combination of hydrogen or deuterium gas with combustion products of ethylene in nitrogen trifluoride. They were invented by George C. Pimentel.

Excimer Lasers

Excimer lasers are powered by a chemical reaction involving an *excited dimer*, or *excimer*, which is a short-lived dimeric or heterodimeric molecule formed from two species (atoms), at least one of which is in an excited electronic state.

They typically produce ultraviolet light, and are used in semiconductor photolithography and in LASIK eye surgery. Commonly used excimer molecules include F_2 (fluorine, emitting at 157 nm), and noble gas compounds (ArF [193 nm], KrCl [222 nm], KrF [248 nm], XeCl [308 nm], and XeF [351 nm]).

Solid-State Lasers

Solid-state laser materials are commonly made by "doping" a crystalline solid host with ions that provide the required energy states. For example, the first working laser was a ruby laser, made from ruby (chromium-doped corundum).

Fig. A 50 W FASOR, Based on a Nd:YAG Laser, used at the Starfire Optical Range

The population inversion is actually maintained in the "dopant", such as chromium or neodymium. Formally, the class of solid-state lasers includes also fibre laser, as the active medium (fibre) is in the solid state. Practically, in the scientific literature, solid-state laser usually means a laser with bulk active medium, while wave-guide lasers are caller fibre lasers.

"Semiconductor lasers" are also solid-state lasers, but in the customary laser terminology, "solid-state laser" excludes semiconductor lasers, which have their own name.

Neodymium is a common "dopant" in various solid-state laser crystals, including yttrium orthovanadate (Nd:YVO_4), yttrium lithium fluoride (Nd:YLF) and yttrium aluminium garnet (Nd:YAG). All these lasers can produce high powers in the infrared spectrum at 1064 nm.

They are used for cutting, welding and marking of metals and other materials, and also in spectroscopy and for pumping dye lasers. These lasers are also commonly frequency doubled, tripled or quadrupled to produce 532 nm (green, visible), 355 nm (UV) and 266 nm (UV) light when those wavelengths are needed.

Ytterbium, holmium, thulium, and erbium are other common "dopants" in solid-state lasers. Ytterbium is used in crystals such as Yb:YAG, Yb:KGW, Yb:KYW, Yb:SYS, Yb:BOYS, Yb:CaF2, typically operating around 1020-1050 nm. They are potentially very efficient and high powered due to a small quantum defect. Extremely high powers in ultrashort pulses can be achieved with Yb:YAG.

Holmium-doped YAG crystals emit at 2097 nm and form an efficient laser operating at infrared wavelengths strongly absorbed by water-bearing tissues. The Ho-YAG is usually operated in a pulsed mode, and passed through optical fibre surgical devices to resurface joints, remove rot from teeth, vaporize cancers, and pulverize kidney and gall stones.

Titanium-doped sapphire (Ti:sapphire) produces a highly tunable infrared laser, commonly used for spectroscopy as well as the most common ultrashort pulse laser.

Thermal limitations in solid-state lasers arise from unconverted pump power that manifests itself as heat and phonon energy. This heat, when coupled with a high thermo-optic coefficient (dn/dT) can give rise to thermal lensing as well as reduced quantum efficiency. These types of issues can be overcome by another novel diode-pumped solid-state laser, the diode-pumped thin disk laser.

The thermal limitations in this laser type are mitigated by utilizing a laser medium geometry in which the thickness is much smaller than the diameter of the pump beam. This allows for a more even thermal gradient in the material. Thin disk lasers have been shown to produce up to kilowatt levels of power.

Fibre-Hosted Lasers

Solid-state lasers where the light is guided due to the total internal reflection in an optical fibre are called fibre lasers. Guiding of light allows extremely long gain regions providing good cooling conditions; fibers have high surface area to volume ratio which allows efficient cooling. In addition, the fiber's waveguiding properties tend to reduce thermal distortion of the beam. Erbium and ytterbium ions are common active species in such lasers. Quite often, the fibre laser is designed as a double-clad fibre. This type of fibre consists of a fibre core, an inner cladding and an outer cladding.

The index of the three concentric layers is chosen so that the fibre core acts as a single-mode fibre for the laser emission while the outer cladding acts as a highly multimode core for the pump laser. This lets the pump propagate a large amount of power into and through the active inner core region, while still having a high numerical aperture (NA) to have easy launching conditions.

Pump light can be used more efficiently by creating a fibre disk laser, or a stack of such lasers.

Fibre lasers have a fundamental limit in that the intensity of the light in the fibre cannot be so high that optical nonlinearities induced by the local electric field strength can become dominant and prevent laser operation and/or lead to the material destruction of the fibre.

This effect is called photodarkening. In bulk laser materials, the cooling is not so efficient, and it is difficult to separate the effects of photodarkening from the thermal effects, but the experiments in fibers show that the photodarkening can be attributed to the formation of long-living color centers.

Photonic Crystal Lasers

Photonic crystal lasers are lasers based on nano-structures that provide the mode confinement and the density of optical states (DOS) structure required for the feedback to take place. They are typical micrometre-sized and tunable on the bands of the photonic crystals.

SEMICONDUCTOR LASERS

Semiconductor lasers are also solid-state lasers but have a different mode of laser operation.

Commercial laser diodes emit at wavelengths from 375 nm to 1800 nm, and wavelengths of over 3 μm have been demonstrated. Low power laser diodes are used in laser printers and CD/DVD players. More powerful laser diodes are frequently used to optically pump other lasers with high efficiency. The highest power industrial laser diodes, with power up to 10 kW (70dBm), are used in industry for cutting and welding.

External-cavity semiconductor lasers have a semiconductor active medium in a larger cavity. These devices can generate high power outputs with good beam quality, wavelength-tunable narrow-linewidth radiation, or ultrashort laser pulses.

Fig. A 5.6 mm 'closed can' Commercial Laser Diode, Probably from a CD or DVD Player

Vertical cavity surface-emitting lasers (VCSELs) are semiconductor lasers whose emission direction is perpendicular to the surface of the wafer. VCSEL devices typically have a more circular output beam than conventional laser diodes, and potentially could be much cheaper to manufacture. As of 2005, only 850 nm VCSELs are widely available, with 1300 nm VCSELs beginning to be commercialized, and 1550 nm devices an area of research.

VECSELs are external-cavity VCSELs. Quantum cascade lasers are semiconductor lasers that have an active transition between energy *sub-bands* of an electron in a structure containing several quantum wells. The development of a silicon laser is important in the field of optical computing, since it means that if silicon, the chief ingredient of computer chips, were able to produce lasers, it would allow the light to be manipulated like electrons are in normal integrated circuits. Thus, photons would replace electrons in the circuits, which dramatically increases the speed of the computer.

Unfortunately, silicon is a difficult lasing material to deal with, since it has certain properties which block lasing. However, recently teams have produced silicon lasers through methods such as fabricating the lasing material from silicon and other semiconductor materials, such as indium(III) phosphide or gallium(III) arsenide, materials which allow coherent light to be produced from silicon. These are called hybrid silicon laser. Another type is a Raman laser, which takes advantage of Raman scattering to produce a laser from materials such as silicon.

Dye Lasers

Dye lasers use an organic dye as the gain medium. The wide gain spectrum of available dyes allows these lasers to be highly tunable, or to produce very short-duration pulses (on the order of a few femtoseconds)

Free Electron Lasers

Free electron lasers, or FELs, generate coherent, high power radiation, that is widely tunable, currently ranging in wavelength from microwaves, through terahertz radiation and infrared, to the visible spectrum, to soft X-rays. They have the widest frequency range of any laser type.

While FEL beams share the same optical traits as other lasers, such as coherent radiation, FEL operation is quite different. Unlike gas, liquid, or solid-state lasers, which rely on bound atomic or molecular states, FELs use a relativistic electron beam as the lasing medium, hence the term *free electron.*

Exotic Laser Media

In September 2007, the BBC News reported that there was speculation about the possibility of using positronium annihilation to drive a very powerful gamma ray laser. Dr. David Cassidy of the University of California, Riverside proposed that a single such laser could be used to ignite a nuclear fusion reaction, replacing the hundreds of lasers used in typical inertial confinement fusion experiments.

Space-based X-ray lasers pumped by a nuclear explosion

have also been proposed as antimissile weapons. Such devices would be one-shot weapons.

USES

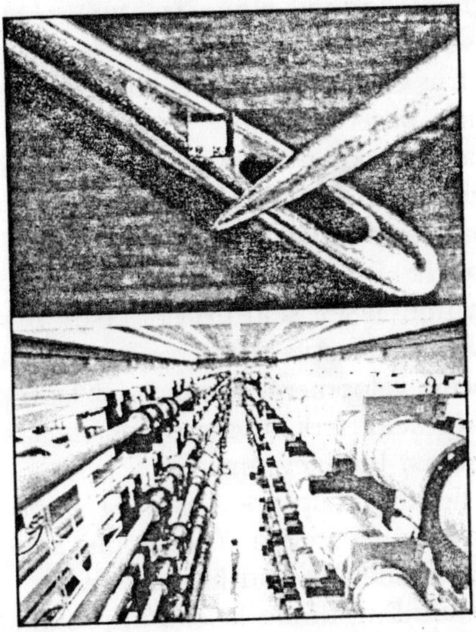

Fig. Lasers Range in size from Microscopic Diode Lasers (top) with Numerous Applications, to Football Field sized Neodymium Glass Lasers (Bottom) used for Inertial Confinement Fusion, Nuclear Weapons Research and other High Energy Density Physics Experiments.

When lasers were invented in 1960, they were called "a solution looking for a problem". Since then, they have become ubiquitous, finding utility in thousands of highly varied applications in every section of modern society, including consumer electronics, information technology, science, medicine, industry, law enforcement, entertainment, and the military.

The first application of lasers visible in the daily lives of the general population was the supermarket barcode scanner, introduced in 1974. The laserdisc player, introduced in 1978, was the first successful consumer product to include a laser, but the compact disc player was the first laser-equipped device to become truly common in consumers' homes, beginning in 1982, followed shortly by laser printers.

Some of the other applications include:

- Medicine: Bloodless surgery, laser healing, surgical treatment, kidney stone treatment, eye treatment, dentistry
- Industry: Cutting, welding, material heat treatment, marking parts
- Defence: Marking targets, guiding munitions, missile defence, electro-optical countermeasures (EOCM), alternative to radar
- Research: Spectroscopy, laser ablation, Laser annealing, laser scattering, laser interferometry, LIDAR, Laser capture microdissection
- Product development/commercial: laser printers, CDs, barcode scanners, thermometers, laser pointers, holograms, bubblegrams.
- Laser lighting displays: Laser light shows
- Laser skin procedures such as acne treatment, cellulite reduction, and hair removal.

In 2004, excluding diode lasers, approximately 131,000 lasers were sold worldwide, with a value of US$2.19 billion. In the same year, approximately 733 million diode lasers, valued at $3.20 billion, were sold.

Examples by Power

Different uses need lasers with different output powers. Lasers that produce a continuous beam or a series of short pulses can be compared on the basis of their average power. Lasers that produce pulses can also be characterized based on the *peak* power of each pulse. The peak power of a pulsed laser is many orders of magnitude greater than its average power. The average output power is always less than the power consumed.

The continuous or average power required for some uses:

- Less than 1 mW - Consumer laser pointers
- 5 mW – CD-ROM drive
- 5–10 mW – DVD player or DVD-ROM drive

- 100 mW - High-speed CD-RW burner
- 250 mW - Consumer DVD-R burner
- 1 W - green laser in current Holographic Versatile Disc prototype development
- 1–20 W - output of the majority of commercially available solid-state lasers used for micro machining
- 30–100 W - typical sealed CO_2 surgical lasers
- 100–3000 W (peak output 1.5 kW) - typical sealed CO_2 lasers used in industrial laser cutting
- 1 kW - Output power expected to be achieved by a prototype 1 cm diode laser bar

Examples of pulsed systems with high peak power:

- 700 TW (700×10^{12} W) - The National Ignition Facility is working on a system that, when complete, will contain a 192-beam, 1.8-megajoule laser system adjoining a 10-meter-diameter target chamber. The system is expected to be completed in April 2009.
- 1.3 PW (1.3×10^{15} W) - world's most powerful laser as of 1998, located at the Lawrence Livermore Laboratory

Hobby Uses

In recent years, some hobbyists have taken interests in lasers. Lasers used by hobbyists are generally of class IIIa or IIIb, although some have made their own class IV types. However, compared to other hobbyists, laser hobbyists are far less common, due to the cost and potential dangers involved. Due to the cost of lasers, some hobbyists use inexpensive means to obtain lasers, such as extracting diodes from DVD burners.

Hobbyists also have been taking surplus pulsed lasers from retired military applications and modifying them for pulsed holography. Pulsed Ruby and Pulsed YAG lasers have been used.

Chapter 13

Laser Safety

Laser safety is the avoidance of laser accidents, especially those involving eye injuries. Since even relatively small amounts of laser light can lead to permanent eye injuries, the sale and usage of lasers is typically subject to government regulations.

Moderate and high-power lasers are potentially hazardous because they can burn the retina of the eye, or even the skin. To control the risk of injury, various specifications, for example ANSI Z136 in the US and IEC 60825 internationally, define "classes" of laser depending on their power and wavelength. These regulations also prescribe required safety measures, such as labeling lasers with specific warnings, and wearing laser safety goggles when operating lasers.

Fig. Warning Symbol for Lasers.

Even the first laser was recognized as being potentially dangerous. Theodore Maiman characterized the first laser as having a power of one "Gillette" as it could burn through one Gillette razor blade. Today, it is accepted that even low-power lasers with only a few milliwatts of output power can be hazardous to human eyesight, when the beam from such a laser hits the eye directly or after reflection from a shiny surface.

At wavelengths which the cornea and the lens can focus well, the coherence and low divergence of laser light means that it can be focused by the eye into an extremely small spot on the retina, resulting in localized burning and permanent damage in seconds or even less time.

Lasers are usually labeled with a safety class number, which identifies how dangerous the laser is:

- Class I/1 is inherently safe, usually because the light is contained in an enclosure, for example in cd players.
- Class II/2 is safe during normal use; the blink reflex of the eye will prevent damage. Usually up to 1 mW power, for example laser pointers.
- Class IIIa/3R lasers are usually up to 5 mW and involve a small risk of eye damage within the time of the blink reflex. Staring into such a beam for several seconds is likely to cause (minor) eye damage.
- Class IIIb/3B can cause immediate severe eye damage upon exposure. Usually lasers up to 500 mW, such as those in cd and dvd burners.
- Class IV/4 lasers can burn skin, and in some cases, even scattered light can cause eye and/or skin damage. Many industrial and scientific lasers are in this class.

The indicated powers are for visible-light, continuous-wave lasers. For pulsed lasers and invisible wavelengths, other power limits apply. People working with class 3B and class 4 lasers can protect their eyes with safety goggles which are designed to absorb light of a particular wavelength.

Certain infrared lasers with wavelengths beyond about 1.4 micrometres are often referred to as being "eye-safe". This is because the intrinsic molecular vibrations of water molecules very strongly absorb light in this part of the spectrum, and thus a laser beam at these wavelengths is attenuated so completely as it passes through the eye's cornea that no light remains to be focused by the lens onto the retina. The label "eye-safe" can be misleading, however, as it only applies to relatively low power continuous wave beams and any high power or Q-switched laser at these wavelengths can burn the cornea, causing severe eye damage.

LASER RADIATION HAZARDS

Laser radiation predominantly causes injury via thermal effects. Even moderately powered lasers can cause injury to the eye. High power lasers can also burn the skin. Some lasers are so powerful that even the diffuse reflection from a surface can be hazardous to the eye.

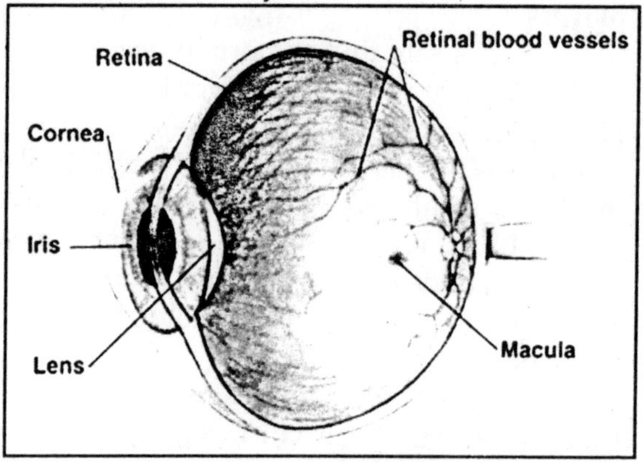

Fig. Diagram of a Human Eye

The coherence, the low divergence angle of laser light and the focusing mechanism of the eye means that laser light can be concentrated into an extremely small spot on the retina. A transient increase of only 10 °C can destroy retinal photoreceptors.

If the laser is sufficiently powerful, permanent damage can occur within a fraction of a second, faster than the blink of an eye. Sufficiently powerful visible to near infrared laser radiation (400-1400 nm) will penetrate the eyeball and may cause heating of the retina, whereas exposure to laser radiation with wavelengths less than 400 nm and greater than 1400 nm are largely absorbed by the cornea and lens, leading to the development of cataracts or burn injuries.

Infrared lasers are particularly hazardous, since the body's protective "blink reflex" response is triggered only by visible light. For example, some people exposed to high power Nd:YAG laser emitting invisible 1064 nm radiation, may not feel pain or notice immediate damage to their eyesight. A pop or click noise emanating from the eyeball may be the only indication that retinal damage has occurred i.e. the retina was heated to over 100 °C resulting in localized explosive boiling accompanied by the immediate creation of a permanent blind spot.

DAMAGE MECHANISMS

Lasers can cause damage in biological tissues, both to the eye and to the skin, due to several mechanisms. Thermal damage, or burn, occurs when tissues are heated to the point where denaturation of proteins occurs. Another mechanism is photochemical damage, where light triggers chemical reactions in tissue.

Photochemical damage occurs mostly with short-wavelength (blue) and ultra-violet light and can be accumulated over the course of hours. Laser pulses shorter than about 1 μs can cause a rapid raise in temperature, resulting in explosive boiling of water. The shock wave from the explosion can subsequently cause damage relatively far away from the point of impact. Ultrashort pulses can also exhibit self-focusing in the transparent parts of the eye, leading to an increase of the damage potential compared to longer pulses with the same energy.

The eye focuses visible and near-infrared light onto the retina. A laser beam can be focused to an intensity on the retina which may be up to 2×10^5 times higher than at the point where the laser beam enters the eye. Most of the light is absorbed by melamine pigments in the pigment epithelium just behind the photoreceptors, and causes burns in the retina.

Ultraviolet light with wavelengths shorter than 400 nm tends to be absorbed in the cornea and lens, where it can produce injuries at relatively low powers due to photochemical damage. Infrared light mainly causes thermal damage to the retina at near-infrared wavelengths and to more frontal parts of the eye at longer wavelengths. The table below summarizes the various medical conditions caused by lasers at different wavelengths, not including injuries due to pulsed lasers.

Wavelength Range	Pathological Effect
180–315 nm (UV-B, UV-C)	photokeratitis (inflammation of the cornea, equivalent to sunburn)
315–400 nm (UV-A)	photochemical cataract (clouding of the eye lens)
400–780 nm (visible)	photochemical damage to the retina, retinal burn
780–1400 nm (near-IR)	cataract, retinal burn
1.4–3.0μm (IR)	aqueous flare (protein in the aqueous humour), cataract, corneal burn
3.0 μm–1 mm	corneal burn

The skin is usually much less sensitive to laser light than the eye, but excessive exposure to ultraviolet light from any source (laser or non-laser) can cause short- and long-term effects similar to sunburn, while visible and infrared wavelengths are mainly harmful due to thermal damage.

LASERS AND AVIATION SAFETY

Since 1990 there have been 400 incidents of lasers directed at aircraft within the United States. These concerns have led to an inquiry in the US congress. Exposure to hand-held laser light under such circumstances may seem trivial given the brevity of exposure, the large distances involved and beam spread of up to several metres. However, laser exposure may create dangerous conditions such as flash blindness.

If this occurs during a critical moment in aircraft operation, the aircraft may be endangered. In addition, some 18% to 35% of the population possess the autosomal dominant genetic trait, photic sneeze, that causes the affected individual to experience an involuntary sneezing fit when exposed to a sudden flash of light. Some observers believe that the danger is greatly exaggerated, at least for small hand-held lasers.

Maximum Permissible Exposure

The *maximum permissible exposure* (MPE) is the highest power or energy density (in W/cm^2 or J/cm^2) of a light source that is considered safe, i.e. that has a negligible probability for creating a damage. It is usually about 10% of the dose that has a 50% chance of creating damage under worst-case conditions. The MPE is measured at the cornea of the human eye or at the skin, for a given wavelength and exposure time.

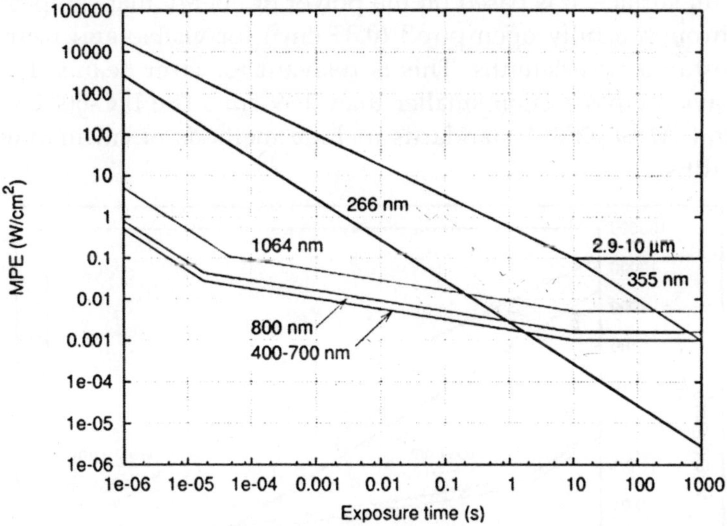

A calculation of the MPE for occular exposure takes into account the various ways light can act upon the eye. For example, deep-ultraviolet light causes accumulating damage, even at very low powers. Infrared light with a wavelength longer than about 1400 nm is absorbed by the transparent parts of the eye before it reaches the retina, which means that the MPE for these wavelengths is higher than for visible light.

In addition to the wavelength and exposure time, the MPE takes into account the spatial distribution of the light (from a laser or otherwise). Collimated laser beams of visible and near-infrared light are especially dangerous at relatively low powers because the lens focuses the light onto a tiny spot on the retina.

Light sources with a smaller degree of spatial coherence than a well-collimated laser beam lead to a distribution of the light over a larger area on the retina. For such sources, the MPE is higher than for collimated laser beams. In the MPE calculation, the worst-case scenario is assumed, in which the eye lens focuses the light into the smallest possible spot size on the retina for the particular wavelength and the pupil is fully open.

Although the MPE is specified as power or energy per unit surface, it is based on the power or energy that can pass through a fully open pupil (0.39 cm^2) for visible and near-infrared wavelengths. This is relevant for laser beams that have a cross-section smaller than 0.39 cm^2. The IEC-60825-1 and ANSI Z136.1 standards include methods of calculating MPEs.

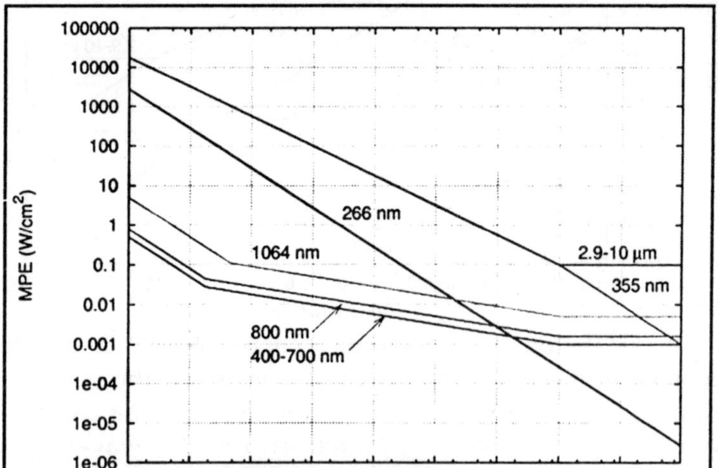

Regulations

In various jurisdictions, standards bodies, legislation, and government regulations define classes of laser according to the risks associated with them, and define required safety measures for people who may be exposed to those lasers.

In the European Community, eye protection requirements are specified in European norm EN 207. In addition to EN 207, European norm EN 208 specifies requirements for goggles for use during beam alignment. These transmit a portion of the laser light, permitting the operator to see where the beam is, and do not provide complete protection against a direct laser beam hit. Finally, European norm EN 60825 specifies optical densities in extreme situations.

In the U.S., guidance for the use of protective eyewear, and other elements of safe laser use, is given in the ANSI Z136 series of standards. They are:

ANSI Z136.1 -	Safe Use of Lasers
ANSI Z136.2 -	Safe Use of Lasers in Optical Fibre Communication Systems Utilizing Laser Diode and LED Sources
ANSI Z136.3 -	Safe Use of Lasers in Health Care Facilities
ANSI Z136.5 -	Safe Use of Lasers in Educational Institutions
ANSI Z136.6 -	Safe Use of Lasers Outdoors

The U.S. Food and Drug Administration (FDA) requires all class IIIb and class IV lasers offered in commerce in the US to have five standard safety features: a key switch, a safety interlock dongle, a power indicator, an aperture shutter, and an emission delay (normally two to three seconds). OEM lasers, designed to be parts of other components (such as DVD burners) are exempt from this requirement. Some non-portable lasers may not have a safety dongle or an emission delay, but have an emergency stop button and/or a remote switch.

Classification

Lasers have been classified by wavelength and maximum output power into four classes and a few subclasses since the early 1970s. The classifications categorize lasers according to their ability to produce damage in exposed people, from class 1 (no hazard during normal use) to class 4 (severe hazard for eyes and skin).

There are two classification systems, the "old system" used before 2002, and the "revised system" being phased in since 2002. The latter reflects the greater knowledge of lasers that has been accumulated since the original classification system was devised, and permits certain types of lasers to be recognized as having a lower hazard than was implied by their placement in the original classification system.

The revised system is part of the revised IEC 60825 standard. From 2007, the revised system is also incorporated into the US-oriented ANSI Laser Safety Standard (ANSI Z136.1). Since 2007, labeling according to the revised system is accepted by the FDA on laser products imported into the US.

The old and revised systems can be distinguished by the 1M, 2M and 3R classes used only in the revised system and the 2A and 3A classes used only in the old system. Class numbers were designated using Roman numerals (I–IV) in the US under the old system and Arabic numerals (1–4) in the EU. The revised system uses Arabic numerals (1–4) in all jurisdictions.

The classification of a laser is based on the concept of *accessible emission limits* (AEL) that are defined for each laser class. This is usually a maximum power (in W) or energy (in J) that can be emitted in a specified wavelength range and exposure time. For infrared wavelengths above 4 μm, it is specified as a maximum power density (in W/m^2).

It is the responsibility of the manufacturer to provide the correct classification of a laser, and to equip the laser with appropriate warning labels and safety measures as prescribed by the regulations. Safety measures used with the more powerful lasers include key-controlled operation, warning lights to indicate laser light emission, a beam stop or attenuator, and an electrical contact that the user can connect to an emergency stop or interlock.

Safety Measures

Most scientists involved with lasers agree on the following minimum guidelines.

- Everyone who touches a laser should be aware of the risks. This awareness is not just a matter of time spent with lasers; to the contrary, long-term dealing with invisible risks (such as from infrared laser beams) tends to reduce risk awareness, rather than to sharpen it.
- Optical experiments are carried out on an optical table, where all laser beams travel in the horizontal plane only, and all beams are stopped at the edges of the table. Users make sure never to put their eyes at the level of the horizontal plane where the beams are, in case a reflected beam leaves the table. The idea is that common mistakes, such as a misplaced mirror, will not cause beams to leave the plane of the table and thus reach the eyes.
- Watches and other jewelry that might enter the optical plane are not allowed in the laboratory. All non-optical objects that are close to the optical plane should have a matte finish in order to prevent specular reflections.
- Adequate eye protection is required for everyone in the room if there is a significant risk for eye injury. Opinions about what constitutes a significant risk vary, though.
- High-intensity beams that can cause fire or skin damage (mainly from class 4 and ultraviolet lasers) and that are not frequently modified should be guided through tubes.
- Alignment of beams and optical components is done at a reduced beam power, which is restored after alignment is complete.

PROTECTIVE EYEWEAR

The use of eye protection when operating lasers of classes 3B and 4 in a manner that may result in eye exposure in excess of the MPE is strongly recommended, and is required in the workplace by the U.S. Occupational Safety and Health Administration.

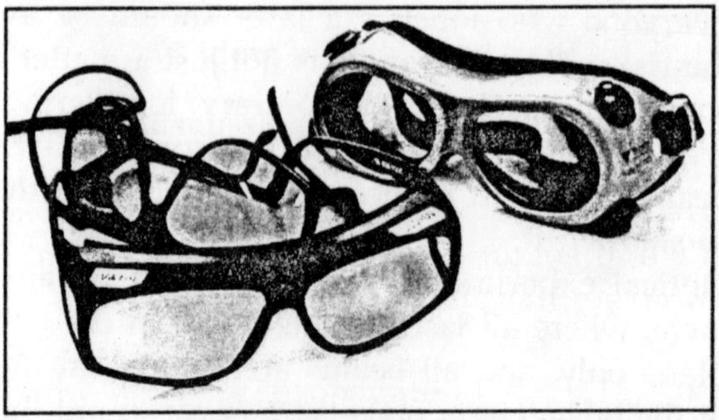

Fig. Laser Goggles

Protective eyewear in the form of spectacles or goggles with appropriately filtering optics can protect the eyes from the reflected or scattered laser light with a hazardous beam power, as well as from direct exposure to a laser beam.

Eyewear must be selected for the specific type of laser, to block or attenuate in the appropriate wavelength range. For example, eyewear absorbing 532 nm typically has an orange appearance, transmitting wavelengths larger than 550 nm. Such eyewear would be useless as protection against a laser emitting at 800 nm.

Eyewear is rated for optical density (OD), which is the base-10 logarithm of the attenuation factor by which the optical filter reduces beam power. For example, eyewear with OD 3 will reduce the beam power in the specified wavelength range by a factor of 1,000. In addition to an optical density sufficient to reduce beam power to below the maximum permissible exposure, laser eyewear used where direct beam exposure is possible should be able to withstand a direct hit from the laser beam without breaking. The protective specifications (wavelengths and optical densities) are usually printed on the goggles, generally near the top of the unit. In the European Community, manufacturers are required by European norm EN 207 to specify the maximum power rating rather than the optical density.

LASER SAFETY OFFICER

In many jurisdictions, organizations that operate lasers

are required to appoint a laser safety officer (LSO). The LSO is responsible for ensuring that safety regulations are followed by all other workers in the organization.

LASER SAFETY IN RESEARCH ENVIRONMENTS

It is common in research in both university and industrial laboratories for operators to violate safety regulations and remove their eye protection during certain procedures, or even to avoid wearing it altogether. Some find the use of safety glasses over a long time to be uncomfortable, and in many types of optical experiments it is also inconvenient. For example in spectroscopy, the experimental arrangement is constantly being modified and fine-tuned.

This requires knowledge of beam location, which is often most simply achieved with the naked eye, although other, often less accurate, detection methods are available. In this situation, many scientists assign a higher priority to convenience and comfort than to safety and regulatory compliance, and routinely breach the laser safety regulations. Sometimes it is perceived as unavoidable when working with, for example, an RGB laser, which would require very careful selection of the Optical Density of protective eyewear or the use of completely black goggles.

Eye protection is often considered uncomfortable because of both reduced vision and physical discomfort. Especially for goggles that protect against visible-light wavelengths, color vision is impaired, which means that it becomes hard to see green or red indicator lights on equipment and to recognize tools. Moreover, many types of goggles transmit less than 30 per cent of visible light, which means that standard work environment lighting levels may be inadequate, leading to increased risks for other accidents, such as tripping over cables. Finally, the weight, fit, and ventilation may cause physical discomfort.

When eye protection is neglected, scientists attempt to minimize risks by other means, such as keeping all beams within a restricted horizontal space, and removing jewelry. However, the reduced safety measures commonplace in scientific environments cannot completely prevent accidents. It is not uncommon for someone who did several years of laser-related research to have experienced an accident that resulted in a small, but permanent eye injury, typically a blind spot somewhere in the peripheral vision.

Experimentalists often feel that safety eyewear is not necessary when dealing with an experiment carried out within the horizontal plane. However, in a non-trivial optical setup, it is very hard to ensure that all mirrors, filters, and lenses are strictly kept in a vertical position at all times, particularly when the setup is constantly modified, and that metallic tools such as screwdrivers also can redirect a beam.

Stray reflections are usually unnoticed until an accident occurs. Since nobody can guarantee that all these hazards can be safely avoided without wearing safety eyewear, when infrared laser beams with non-negligible powers are used, working without safety eyewear is not permitted by any official safety regulations.

Laser Pointers

In the period from 2000 to 2008, increasing attention has been paid to the risks posed by so called laser pointers and laser pens. Typically, sales of laser pointers is restricted to either class 3A (<5 mW) or class 2 (<1 mW), depending on local regulations.

For example, in the US, class 3A is the maximum permitted, unless a key actuated control or other safety features are provided and in the UK, class 2 is the maximum allowed class. However, because enforcement is often not very strict, class 3A laser pointers are often available for sale even in countries where they are not allowed.

Van Norren et al. could not find a single example in the medical literature of a <1 mW class II laser causing eyesight damage. Mainster et al. provide one case, an 11 year old child who temporarily damaged her eyesight by holding an approximately 5 mW red laser pointer close to the eye and staring into the beam for 10 seconds, she experienced scotoma (a blind spot) but fully recovered after 3 months.

Luttrulla & Hallisey describe a similar case, a 34 year old male who stared into the beam of a class IIIa red laser for 30 to 60 seconds, causing temporary central scotoma and visual field loss. His eyesight fully recovered within 2 days, at the time of his eye exam. An intravenous fundus fluorescein angiogram, a technique used by ophthalmologists to visualise the retina of the eye in fine detail, identified subtle discoloration of the fovea.

Thus, it appears that a brief 0.25-second exposure to a <5 mW laser such as found in red laser pointers does not pose a threat to eye health. On the other hand there is a potential for injury if a person deliberately stares into a beam of a class IIIa laser for few seconds or more at close range. Even if injury occurs, most people will fully recover their vision. Further experienced discomforts than these may be psychological rather than physical.

With regard to green laser pointers the safe exposure time may be less, and with even higher powered lasers instant permanent damage should be expected. These conclusions must be qualified with recent theoretical observations that certain prescription medications may interact with some wavelengths of laser light, causing increased sensitivity (phototoxicity).

Beyond the question of physical injury to the eye from a laser pointer, several other undesirable effects are possible. These include short-lived flash blindness if the beam is encountered in darkened surroundings, as when driving at night. This may result in momentary loss of vehicular control. Lasers pointed at aircraft are a hazard to aviation. A police officer seeing a red dot on his chest may conclude that a sniper is targeting him and take aggressive action.

In addition, the startle reflex exhibited by some exposed unexpectedly to laser light of this sort has been reported to have resulted in cases of self-injury or loss of control. For these and similar reasons, the US Food and Drug Administration has advised that laser pointers are not toys and should not be used by minors except under the direct supervision of an adult.

Non-Beam Hazards - Electrical and Other

A discussion of laser safety would not be complete without mention of non-beam hazards that are often associated with use of laser systems. Many lasers are high voltage devices, typically 400 V upward for a small 5 mJ pulsed laser, and exceeding many kilovolts in higher powered lasers. This, coupled with high pressure water for cooling the laser and other associated electrical equipment can create a greater hazard than the laser beam itself.

Electric equipment should generally be installed at least 250 mm / 10 inches above the floor to reduce electric risk in the case of flooding. Optical tables, lasers, and other equipment should be well grounded. Enclosure interlocks should be respected and special precautions taken during troubleshooting.

In addition to the electrical hazards, lasers may create chemical, mechanical, and other hazards specific to particular installations. Chemical hazards may include materials intrinsic to the laser, such as beryllium oxide in argon ion laser tubes, halogens in excimer lasers, organic dyes dissolved in toxic or flammable solvents in dye lasers, and heavy metal vapors and asbestos insulation in helium cadmium lasers.

They may also include materials released during laser processing, such as metal fumes from cutting or surface treatments of metals or the complex mix of decomposition products produced in the high energy plasma of a laser cutting plastics.

Mechanical hazards may include moving parts in vacuum and pressure pumps; implosion or explosion of flashlamps, plasma tubes, water jackets, and gas handling equipment.

High temperatures and fire hazards may also result from the operation of high-powered Class IIIB or any Class IV Laser.

In commercial laser systems, hazard mitigations such as the presence of fusible plugs, thermal interrupters, and pressure relief valves reduce the hazard of, for example, a steam explosion arising from an obstructed water cooling jacket. Interlocks, shutters, and warning lights are often critical elements of modern commercial installations. In older lasers, experimental and hobby systems, and those removed from other equipment (OEM units) special care must be taken to anticipate and reduce the consequences of misuse as well as various failure modes.

Lasers as Weapons

Laser beams are famously employed as weapon systems in science fiction, but actual laser weapons are only beginning to enter the market. The general idea of laser-beam weaponry is to hit a target with a train of brief pulses of light. The rapid evaporation and expansion of the surface causes shockwaves that damage the target.

The power needed to project a high-powered laser beam of this kind is difficult for current mobile power technology. Public prototypes are chemically-powered gas dynamic lasers.

Lasers of all but the lowest powers can potentially be used as incapacitating weapons, through their ability to produce temporary or permanent vision loss in varying degrees when aimed at the eyes. The degree, character, and duration of vision impairment caused by eye exposure to laser light varies with the power of the laser, the wavelength(s), the collimation of the beam, the exact orientation of the beam, and the duration of exposure.

Lasers of even a fraction of a watt in power can produce immediate, permanent vision loss under certain conditions, making such lasers potential non-lethal but incapacitating weapons. The extreme handicap that laser-induced blindness represents makes the use of lasers even as non-lethal weapons morally controversial.

In the field of aviation, the hazards of exposure to ground-based lasers deliberately aimed at pilots have grown to the extent that aviation authorities have special procedures to deal with such hazards.

LASER APPLICATIONS

In manufacturing, lasers are used for cutting, bending, and welding metal and other materials, and for "marking"—producing visible patterns such as letters by changing the properties of a material or by inscribing its surface. In science, lasers are used for many applications.

One of the more common is laser spectroscopy, which typically takes advantage of the laser's well-defined wavelength or the possibility of generating very short pulses of light. Lasers are used by the military for range-finding, target designation, and illumination. Lasers have also begun to be tested for directed-energy weapons. Lasers are used in medicine for surgery, diagnostics, and therapeutic applications.

There are many scientific, military, medical and commercial laser applications which have been developed since the invention of the laser in the 1958. The coherency, high monochromaticity, and ability to reach extremely high powers are all properties which allow for these specialized applications.

SPECTROSCOPY

Most types of laser are an inherently pure source of light; they emit near-monochromatic light with a very well defined range of wavelengths. By careful design of the laser components, the purity of the laser light (measured as the "linewidth") can be improved more than the purity of any other light source. This makes the laser a very useful source for spectroscopy.

The high intensity of light that can be achieved in a small, well collimated beam can also be used to induce a nonlinear optical effect in a sample, which makes techniques such as Raman spectroscopy possible. Other spectroscopic techniques based on lasers can be used to make extremely sensitive detectors of various molecules, able to measure molecular concentrations in the parts-per-trillion (ppt) level. Due to the high power densities achievable by lasers, beam-induced atomic emission is possible: this technique is termed Laser induced breakdown spectroscopy (LIBS).

Lasers may also be indirectly used in spectroscopy as a micro-sampling system, a technique termed Laser ablation (LA), which is typically applied to ICP-MS apparatus resulting in the powerful LA-ICP-MS.

The principles of laser spectroscopy are discussed by

Demtroder and the use of tunable lasers in spectroscopy are described in Tunable Laser Applications.

Lunar Laser Ranging

When the Apollo astronauts visited the moon, they planted retroreflector arrays to make possible the Lunar Laser Ranging Experiment. Laser beams are focused through large telescopes on Earth aimed toward the arrays, and the time taken for the beam to be reflected back to Earth measured to determine the distance between the Earth and Moon with high precision.

Material Processing

Laser cutting, laser welding, laser brazing, laser bending, laser engraving or marking, laser cleaning, weapons etc.

Photochemistry

Some laser systems, through the process of modelocking, can produce extremely brief pulses of light - as short as picoseconds or femtoseconds (10^{-12} - 10^{-15} seconds). Such pulses can be used to initiate and analyse chemical reactions, a technique known as *photochemistry*. The short pulses can be used to probe the process of the reaction at a very high temporal resolution, allowing the detection of short-lived intermediate molecules. This method is particularly useful in biochemistry, where it is used to analyse details of protein folding and function.

Laser Cooling

A technique that has had recent success is *laser cooling*. This involves atom trapping, a method where a number of atoms are confined in a specially shaped arrangement of electric and magnetic fields.

Shining particular wavelengths of laser light at the ions or atoms slows them down, thus *cooling* them. As this process is continued, they all are slowed and have the same energy level, forming an unusual arrangement of matter known as a Bose-Einstein condensate.

Nuclear Fusion

Some of the world's most powerful and complex arrangements of multiple lasers and optical amplifiers are used to produce extremely high intensity pulses of light of extremely short duration. These pulses are arranged such that they impact pellets of tritium-deuterium simultaneously from all directions, hoping that the squeezing effect of the impacts will induce atomic fusion in the pellets.

This technique, known as "inertial confinement fusion", so far has not been able to achieve "breakeven", that is, so far the fusion reaction generates less power than is used to power the lasers, but research continues.

Microscopy

Confocal laser scanning microscopy and Two-photon excitation microscopy make use of lasers to obtain blur-free images of thick specimens at various depths. Laser capture microdissection use lasers to procure specific cell populations from a tissue section under microscopic visualization.

Additional laser microscopy techniques include harmonic microscopy, four-wave mixing microscopy and interferometric microscopy.

Military

Military uses of lasers include applications such as target designation and ranging, defensive countermeasures, communications and directed energy weapons. Directed energy weapons such as Boeing's Airborne Laser which can be mounted on a 747 jet is able to burn the skin off enemy missiles.

DEFENSIVE COUNTERMEASURES

Defensive countermeasure applications can range from compact, low power infrared countermeasures to high power, airborne laser systems. IR countermeasure systems use lasers to confuse the seeker heads on heat-seeking anti-aircraft missiles. High power boost-phase intercept laser systems use a complex system of lasers to find, track and destroy intercontinental ballistic missiles.

In this type of system a chemical laser, one in which the laser operation is powered by an energetic chemical reaction, is used as the main weapon beam. The Mobile Tactical High-Energy Laser (MTHEL) is another defensive laser system under development; this is envisioned as a field-deployable weapon system able to track incoming artillery projectiles and cruise missiles by radar and destroy them with a powerful deuterium fluoride laser.

Another example of direct use of a laser as a defensive weapon was researched for the Strategic Defence Initiative (SDI, nicknamed "Star Wars"), and its successor programs. This project would use ground-based or space-based laser systems to destroy incoming intercontinental ballistic missiles (ICBMs). The practical problems of using and aiming these systems were many; particularly the problem of destroying ICBMs at the most opportune moment, the *boost phase* just after launch.

This would involve directing a laser through a large distance in the atmosphere, which, due to optical scattering and refraction, would bend and distort the laser beam, complicating the aiming of the laser and reducing its efficiency.

Another idea to come from the SDI project was the *nuclear-pumped X-ray laser.* This was essentially an orbiting atomic bomb, surrounded by laser media in the form of glass rods; when the bomb exploded, the rods would be bombarded with highly-energetic gamma-ray photons, causing spontaneous and stimulated emission of X-ray photons in the atoms making up the rods.

This would lead to optical amplification of the X-ray photons, producing an X-ray laser beam that would be minimally affected by atmospheric distortion and capable of destroying ICBMs in flight. The *X-ray laser* would be a strictly one-shot device, destroying itself on activation. Some initial tests of this concept were performed with underground nuclear testing; however, the results were not encouraging. Research into this approach to missile defence was discontinued after the SDI programme was cancelled.

The United States Air Force has experimented with using

lasers combined with high-altitude airships as a potential means for a missile defence shield but also as a means to destroy enemy spacecraft or satellites in low-earth orbit.

TARGETING

Laser Range Finder

A laser rangefinder is a device which uses a laser beam in order to determine the distance to a reflective object. The most common form of laser rangefinder operates on the time of flight principle by sending a laser pulse in a narrow beam towards the object and measuring the time taken by the pulse to be reflected off the target and returned to the sender.

Due to the high speed of light, this technique is not appropriate for high precision sub-millimeter measurements, where triangulation and other techniques are often used.

Fig. A long range laser rangefinder LRB20000 is capable of measuring distance up to 20 km; mounted on a tripod with an angular mount. The resulting system also provides azimuth and elevation measurements

A *laser rangefinder* is a device consisting of a pulsed laser and a light detector. By measuring the time taken for light to reflect off a far object, and knowing the speed of light, the range to the object can be found. A laser rangefinder is thus a simple form of *LIDAR*. The distance to the target can then be used to aim a weapon such as a tank's main gun.

Operation

Pulse

The pulse may be coded in order to reduce the chance that the rangefinder can be jammed. It is possible to use Doppler effect techniques to judge whether the object is moving towards or away from the rangefinder, and if so how fast.

The accuracy of the instrument is determined by the brevity of the laser pulse and the speed of the receiver. One that uses very short, sharp laser pulses and has a very fast detector can range an object to within a few centimeters.

Range

Despite the beam being narrow, it eventually spreads over long distances due to the divergence of the laser beam, as well as to scintillation and beam wander effects, caused by the presence of air bubbles in the air acting as lenses ranging in size from microscopic to roughly half the height of the laser beam's path above the earth. These atmospheric distortions coupled with the divergence of the laser itself and with transverse winds that serve to push the atmospheric heat bubbles laterally may combine to make it difficult to get an accurate reading of the distance of an object, say, beneath some trees or behind bushes, or even over long distances of more than 1 km in open and unobscured desert terrain.

Some of the laser light might reflect off leaves or branches which are closer than the object, giving an early return and a reading which is too low. Alternatively, over distances longer than 1200 ft (365 m), the target, if in proximity to the earth, may simply vanish into a mirage, caused by temperature gradients in the air in proximity to the heated desert bending the laser light. All these effects have to be taken into account.

Discrimination

Some instruments are able to determine multiple returns.

These instruments use waveform-resolving detectors, which means they detect the amount of light returned over a certain time, usually very short. The waveform from a laser pulse that hits a tree and then the ground would have two peaks. The first peak would be the distance to the tree, and the second would be the distance to the ground.

The ability for aircraft-mounted instruments to see "through" dense canopies and other semi-reflective surface such as the ocean provide many applications for airborne instruments such as:

- Creating "bare earth" topographic maps - removing all trees
- Creating vegetation thickness maps
- Bathymetry (measuring topography under the ocean)
- Forest fire hazard
- Overwash threat in barrier islands

TECHNOLOGIES

Time of flight - this measures the time taken for a light pulse to travel to the target and back. With the speed of light known, and an accurate measurement of the time taken, the distance can be calculated. Many pulses are fired sequentially and the average response is most commonly used. This technique requires very accurate sub-nanosecond timing circuitry.

Multiple frequency phase-shift - this measures the phase shift of multiple frequencies on reflection then solves some simultaneous equations to give a final measure.

Interferometry - the most accurate and most useful technique for measuring changes in distance rather than absolute distances.

APPLICATIONS

Military

Rangefinders provide an exact distance to targets located beyond the distance of point-blank shooting to snipers and artillery. They can also be used for military reconciliation and engineering.

Handheld military rangefinders operate at ranges of 2 km up to 25 km and are combined with binoculars or monoculars. When the rangefinder is equipped with a compass and inclinometer it is capable of providing magnetic azimuth, inclination, and height (length) of targets. Some rangefinders can also measure a target's speed in relation to the observer.

Some rangefinders have cable or wireless interfaces to enable them to transfer their measurement(s) data to other equipment like fire control computers. Some models also offer the possibility to use add-on night vision modules. Handheld rangefinders use standard or non-magnetic batteries. The more powerful models of rangefinders measure distance up to 25 km and are normally installed either on a tripod or directly on a vehicle or gun platform. In the latter case rangefinder module is integrated with on-board thermal, night vision and daytime observation equipment. The most advanced military rangefinders can be integrated with computer.

In order to make laser rangefinders and laser-guided weapons less useful against military targets, various military arms may have developed laser-absorbing paint for their vehicles. Regardless, some objects don't reflect laser light very well and using a laser rangefinder on them is difficult.

3-D Modelling

This lidar scanner may be used to scan buildings, rock formations, etc., to produce a 3D model. The lidar can aim its laser beam in a wide range: its head rotates horizontally, a mirror flips vertically. The laser beam is used to measure the distance to the first object on its path.

Laser rangefinders are used extensively in 3-D object recognition, 3-D object modelling, and a wide variety of computer vision-related fields. This technology constitutes the heart of the so-called *time-of-flight* 3D scanners. In contrast to the military instruments, laser rangefinders offer high-precision scanning abilities, with either single-face or 360-degree scanning modes. A number of algorithms have been developed to merge the range data retrieved from multiple angles of a single object in order to produce complete 3-D models with as little error as possible.

One of the advantages that laser rangefinders offer over other methods of computer vision is that the computer does not need to correlate features from two images to determine depth information as in stereoscopic methods.

The laser rangefinders used in computer vision applications often have depth resolutions of tenths of millimeters or less. This can be achieved by using triangulation or refraction measurement techniques as opposed to the time of flight techniques used in LIDAR.

Handheld Distance Meters

Opti-Logic Corporation introduced the first consumer level time of flight handheld laser distance meter in 1987. The original handheld consumer priced laser rangefinders were used for golf. Since that time numerous applications have developed. The most popular use is for hunting.

Until 1993, phase shift instruments were reserved to professional users giving their prices and rich functions like bluetooth measurements transmission. Less expensive models around 150$/€ are emerging from Bosch with the DLE 50 or Stanley Works with the TLM 100.

Sports

Laser rangefinders may be effectively used in various sports that require precision distance measurement, such as golfing, hunting, archery, sniper shooting. Several manufacturers, such as Opti-logic Corporation, Bushnell, Leica, Nikon and Swarovski supply handheld rangefinders at prices well in line with the rest of sports equipment.

Handheld rangefinders are marked in terms of distance they measure to a reflective target, called rated distance. As a rule of thumb, rated distance should be divided by three to estimate actual average distance to a shooting target or golf pin. Practically speaking, a rangefinder marked as XYZ-1000 will give consistent measurements up to 400 m.

TARGET DESIGNATOR

Fig. A Target Designator

Another military use of lasers is as a *laser target designator*. This is a low-power laser pointer used to indicate a target for a precision-guided munition, typically launched from an aircraft. The guided munition adjusts its flight-path to home in to the laser light reflected by the target, enabling a great precision in aiming.

The beam of the laser target designator is set to a pulse rate that matches that set on the guided munition to ensure munitions strike their designated targets and do not follow other laser beams which may be in use in the area. The laser designator can be shone onto the target by an aircraft or nearby infantry. Lasers used for this purpose are usually infrared lasers, so the enemy cannot easily detect the guiding laser light.

FIREARMS

FTP Laser Sight

The laser has in most firearms applications has been used as a tool to enhance the targeting of other weapon systems. For example, a *laser sight* is a small, usually visible-light laser placed on a handgun or rifle aligned to emit a beam parallel to the barrel. Since a laser beam by definition has low divergence, the laser light appears as a small spot even at long distances;

the user places the spot on the desired target and the barrel of the gun is aligned (but not necessarily allowing for windage and the target moving while the bullet travels).

Fig. Revolver Equipped with Laser Sight

Most laser sights use a red laser diode. Others use an infrared diode to produce a dot invisible to the naked human eye but detectable with night vision devices.

The firearms adaptive target acquisition module LLM01 laser light module combines visible and infrared laser diodes. In the late 1990s, green diode pumped solid state laser (DPSS) laser sights (532 nm) became available. Modern laser sights are small and light enough for attachment to the firearms.

In 2007, LaserMax, a company specializing in manufacturing lasers for military and police firearms, introduced the first mass-production green laser available for small arms.

This laser mounts to the underside of a handgun or long arm on the accessory rail. The green laser is supposed to be more visible than the red laser in bright lighting conditions because, for the same wattage, green light appears brighter than red light.

Eye-Targeted Lasers

A non-lethal laser weapon was developed by the U.S. Air Force to temporarily impair an adversary's ability to fire a weapon or to otherwise threaten enemy forces. This unit illuminates an opponent with harmless low-power laser light and can have the effect of dazzling or disorienting the subject or causing him to flee. There remains the possibility of using lasers to blind, since this requires much lower power levels, and is easily achievable in a man-portable unit. However, most nations regard the deliberate permanent blinding of the enemy as forbidden by the rules of war.

In addition to the applications that crossover with military applications, a widely known law enforcement use of lasers is for lidar to measure the speed of vehicles.

Index